WET WEATHER DESIGN AND OPERATION IN WATER RESOURCE RECOVERY FACILITIES

A Special Publication

2014

Water Environment Federation
601 Wythe Street
Alexandria, VA 22314-1994 USA
http://www.wef.org

Wet Weather Design and Operation in Water Resource Recovery Facilities

ISBN 978-1-57278-304-1

About WEF

Founded in 1928, the Water Environment Federation (WEF) is a not-for-profit technical and educational organization of 36,000 individual members and 75 affiliated Member Associations representing water quality professionals around the world. WEF members, Member Associations, and staff proudly work to achieve our mission to provide bold leadership, champion innovation, connect water professionals, and leverage knowledge to support clean and safe water worldwide. To learn more, visit www.wef.org.

For information on membership, publications, and conferences, contact

Water Environment Federation
601 Wythe Street
Alexandria, VA 22314-1994 USA
(703) 684-2400
http://www.wef.org

Prepared by the **Wet Weather Design and Operation in Water Resource Recovery Facilities** Task Force of the **Water Environment Federation.**

Christopher W. Tabor, P.E., *Co-Chair*
Julian Sandino, Ph.D., P.E., BCEE, *Co-Chair*

Virgil C. Adderley, P.E.
Rich Atoulikian, PMP, BCEE, P.E.
Dempsey Ballou, P.E.
Lorenzo Benedetti, Ph.D.
Annie Blissit, E.I.T.
Randall S. Booker, Jr., Ph.D., P.E.
Patrick J. Bradley
Nicholas J. Bucurel, P.E.
Chein-Chi Chang, Ph.D., P.E.
Mark Coleman
Jon Cooper
James Cramer, EI
Matt Crow, P.E.
Scott Cummings
John R. Dening, P.E., CFM
Amanda Dobbs, P.E.
James S. Drake, P.E., BCEE
Cary R. Duchene, P.E.
William Eleazer, P.E.
Richard Finger
James D. Fitzpatrick, P.E.
W. James Gellner, P.E.
Gregory R. Heath, P.E.
Jared C. Hutchins, P.E., ENV SP
Sidney Innerebner, Ph.D., P.E., CWP
Michael D. Jankowski, P.E.
Lawrence P. Jaworski, P.E.; BCEE
James Kerrigan, P.E.
Dale E. Kocarek, P.E., BCEE
Hua (Larry) Li, Ph.D., P.E.
Thomas A. Lyon, P.E.
Jane McLamarrah, Ph.D., P.E.
Henryk Melcer, Ph.D., P.E.
Susan Moisio
Tim O'Brien, P.E.
C. Robert O'Bryan, Jr., P.E.
Carl D. Parrott, P.E.
Stacy J. Passaro, P.E., BCEE
Jurek Patoczka, Ph.D., P.E.
Scott Phipps, P.E.
Joseph C. Reichenberger P.E., BCEE
Nalin Sahni
Jim Scholl
Alan Scrivner, P.E.
Vamsi Seeta
Allen P. Sehloff, P.E.
Siddhartha Sengupta, P.E.
John Shaw, P.E.
John Siczka, P.E.
Viraj de Silva, Ph.D., P.E., BCEE
Stephanie Spalding
Mark Stirrup
Eric Teittinen
David Terrill, P.E.
Peter A. Vanrolleghem, Ph.D., ing.
Miguel Vera
Joseph Viciere, P.E., BCEE
Kevin Weiss
Nancy J. Wheatley
G. Elliott Whitby, Ph.D.
Jann Yamauchi

Under the Direction of the **Municipal Design Subcommittee** of the **Technical Practice Committee**

2014

Water Environment Federation
601 Wythe Street
Alexandria, VA 22314-1994 USA
http://www.wef.org

Special Publications of the Water Environment Federation

The WEF Technical Practice Committee (formerly the Committee on Sewage and Industrial Wastes Practice of the Federation of Sewage and Industrial Wastes Associations) was created by the Federation Board of Control on October 11, 1941. The primary function of the Committee is to originate and produce, through appropriate subcommittees, special publications dealing with technical aspects of the broad interests of the Federation. These publications are intended to provide background information through a review of technical practices and detailed procedures that research and experience have shown to be functional and practical.

Contents

List of Figures

List of Tables

Preface

The purpose of this publication is to provide professionals involved in the design and operations of water resource recovery facilities (WRRFs) with a comprehensive reference of current design and operational practices for dealing with the unique challenges associated with the management of wet weather flows at these facilities.

This publication is intended to be a companion to *Design of Municipal Wastewater Treatment Plants*, 5th Edition (MOP 8) (WEF et al., 2010), and is focused on providing more wet weather flow management perspective to the subject matters covered by this widely relied upon design practitioner resource. However, publication goes beyond MOP 8 by including operational-oriented considerations related to wet weather events. The document comprises 16 chapters that are grouped around planning and configuration for wet weather events in WRRFs and facility processes.

This Special Publication was produced under the direction of Christopher W. Tabor, P.E., Co-Chair, and, Julian Sandino, Ph.D., P.E., BCEE, CH2M HILL, Co-Chair.

In addition to the WEF Task Force and Technical Practice Committee Control Group members, reviewers include Karabo Nthethe and Koos Wiken, Pr. Eng.

Authors' and reviewers' efforts were supported by the following organizations:

AECOM, Los Angeles, California; and Wakefield, Massachusetts
Black & Veatch, Cincinnati, Ohio; Columbus, Ohio; Kansas City, Missouri; St. Louis, Missouri; and Toledo, Ohio
Brentwood Industries, Inc., Reading, Pennsylvania
Brown and Caldwell, Walnut Creek, California
Calgon Carbon Corporation UV Technologies Division, Coraopolis, Pennsylvania
CDM Smith, Cambridge, Massachusetts
CH2M Hill, Cincinnati, Ohio; Cleveland, Ohio; Milwaukee, Wisconsin; and Overland Park, Kansas
City of Houston, Texas
City of Portland, Bureau of Environmental Services, Portland, Oregon,
District of Columbia Water and Sewer Authority, Washington, D.C.
Fox River Water Reclamation District, Elgin, Illinois

Gresham, Smith and Partners, Alpharetta, Georgia; Cincinnati, Ohio; and Nashville, Tennessee
Hatch Mott MacDonald, Iselin, New Jersey
Hazen and Sawyer, P.C., Cincinnati, Ohio; Columbus, Ohio; and Richmond, Virginia
HDR Engineering Inc., Cleveland, Ohio; and Norfolk, Virginia
HRM Environmental, Cedar Park, Texas
Indigo Water Group, Littleton, Colorado
John Shaw Consulting, LLC
LimnoTech, Washington, D.C.
Loyola Marymount University, Los Angeles, California
Malcolm Pirnie, Cleveland, Ohio
Manatee County Public Works Department, Bradenton, Florida
model*EAU* – Université Laval, Québec, Canada
MWH, Broomfield, Colorado
Oklahoma DEQ, Oklahoma City, Oklahoma
Passaro Engineering LLC, Mount Airy, Maryland
Professional Engineering Consultants, P.A., Wichita, Kansas
Stantec, Columbus, Ohio
Trinity River Authority of Texas, Texas
U.S. Environmental Protection Agency, Washington, D.C.
Wade Trim, Detroit, Michigan
WATERWAYS srl, Impruneta, Italy
Woodard & Curran, Portland, Maine

1

Introduction

Christopher W. Tabor, P.E., and
Julian Sandino, Ph.D., P.E., BCEE, CH2M HILL, Co-Chair

1.0 BACKGROUND AND PURPOSE

Wastewater utilities are coming under increasing pressure from regulatory and public realms to manage wet weather conditions more effectively. The flows and accompanying loads associated with these wet weather conditions, which result from combined sewer systems or even separate systems that have significant amounts of infiltration and inflow, can, in many instances, exceed the capacity of an existing water resource recovery facility (WRRF). However, the overall approach necessary to identify, evaluate, and eventually select best management practices (BMPs) for wet weather flows in terms of design and operational practices related to WRRFs is one that contrasts significantly from that typically used by the same utilities in dealing with dry weather treatment needs.

The more predictable and stable nature of dry weather flow conditions allows for the design and operation of treatment facilities to reliably meet treatment requirements expressed as numerical limits as defined in typical National Pollutant Discharge Elimination System permits. This is because short-term (daily, weekly) and longer-term (monthly, seasonal) fluctuations in dry weather influent flows and loads tend to be largely well defined and predictable. This allows the designer to establish treatment solutions for which performance can be anticipated with a higher level of confidence. This is in sharp contrast to the randomly variable (in terms of frequency and duration) and typically higher magnitude of flow and load fluctuations associated with wet weather conditions. As a designer, it is difficult to establish cost-effective facility requirements if the goal is to meet particular numerical limits for a given wet weather event (regardless of whether they are expressed as treated effluent concentrations and/or removal efficiencies) on the basis of a highly variable influent condition with an undetermined duration.

For facility operators used to anticipating and proactively implementing effective operational strategies to deal with the predictable nature of dry weather conditions, wet weather events can become a challenge. These events require operators to continuously monitor changing influent conditions and the corresponding response of the treatment facilities. This, in turn, typically dictates the frequent modification of operational strategies and corresponding facility set points during a single event, increasing the likelihood of not being able to reliably meet static numerical-based treated effluent requirements. Addressing this challenge requires simple, resilient, and reliable treatment technologies to manage wet weather flows that can be appropriately designed and effectively operated to meet technology- and BMP-based limits for these conditions.

The purpose of this publication is to provide professionals involved in the design and operation of WRRFs with a comprehensive reference of current design and operational practices for dealing with the unique challenges associated with the management of wet weather flows at these facilities. Additional information regarding strategic planning and decision-making tools related to wet weather system performance in the context of a regulatory framework is available in the Water Environment Federation (WEF) *Guide for Municipal Wet Weather Strategies* (2013). The material in this publication reflects the experience of WEF's volunteer authors and reviewers from around the world.

2.0 SCOPE AND ORGANIZATION

This publication is intended to be a companion to *Design of Municipal Wastewater Treatment Plants*, 5th Edition (MOP 8) (WEF et al., 2010) and is focused on providing a more wet weather flow management perspective to the subject matters covered so well by this widely relied upon design practitioner resource. However, this publication is intended to also go beyond MOP 8 by including operational-oriented considerations related to wet weather events. The document comprises 16 chapters, grouped around the following two general topics:

- Planning and configuration for wet weather events in WRRFs—included here are topics such as overall design considerations, principles of integrated facility design, site development and facility arrangement, facility hydraulics and pumping, occupational health and safety, support systems, holistic planning and operation (considering collection system operational effects), and principles of system modeling; and

- Facility processes—in general, this publication follows a similar breakdown of facility unit processes as defined in MOP 8. These include flow equalization, preliminary treatment, primary treatment, biological treatment, physical and chemical treatment, disinfection, and residuals.

3.0 REFERENCE

Water Environment Federation; American Society of Civil Engineers; Environmental & Water Resources Institute (2010) *Design of Municipal Wastewater Treatment Plants,* 5th ed.; WEF Manual of Practice No. 8; ASCE Manuals and Reports on Engineering Practice No. 76; Water Environment Federation: Alexandria, Virginia.

2

Overall Design Considerations

Dale E. Kocarek, P.E., BCEE; Patrick J. Bradley; Nicholas J. Bucurel, P.E.; and Jon Cooper

1.0 WATER QUALITY CONSIDERATIONS

1.1 Introduction to Water Quality Standards

Water quality standards are set by the states and reviewed and approved by the U.S. Environmental Protection Agency (U.S. EPA). Clean Water Act (CWA) section 303(c) sets the minimum requirements for states in setting water quality standards. The regulations for implementing section 303 are in 40 *CFR* Part 131. The states have considerable flexibility regarding development and implementation of water quality standards, including both high and low flow scenarios. The water quality standards are used when assessing the status of a waterbody and to determine effects of discharges and other pollutant sources on the water. Additional information about water quality standards and implementation of water quality standards can be found on U.S. EPA's Web sites (http://water.epa.gov/learn/training/standardsacademy/ and http://cfpub.epa.gov/npdes/home.cfm?program_id=45).

1.2 Effects of Untreated Wastewater

1.2.1 Bacteriological Effects

Untreated or partially treated raw wastewater from wet weather discharges through sanitary sewer overflows (SSOs) and combined sewer overflows (CSOs) have the potential to affect the aquatic and terrestrial environment. The term, *bacteria*, includes a broad range of different types of microorganisms. Bacteria of perhaps the most interest to water resource recovery facilities (WRRFs) and the public are those that are considered harmful (pathogenic) to humans. Currently, methods and standards for identifying every type of potentially harmful bacteria or other microorganism do not exist. Historically, these pathogens were measured through an indicator pollutant, fecal coliform, and more recently by the indicator bacteria, *Escherichia coli* and enterococci (see http://water.epa.gov/scitech/swguidance/standards/criteria/health/recreation/index.cfm).

1.2.2 Oxygen Depletion

Dissolved oxygen concentrations in surface waters can range from barely measurable to approximately 10 ppm. In surface waters, dissolved oxygen concentrations below 3 ppm will stress fish and concentrations below 2 ppm will kill some species (depending on length of exposure and species). Fish, similar to other species that are stressed as a result of oxygen depletion, will become susceptible to disease. Increased river flow from wet weather events can resuspend sediments and nutrients and cause dissolved oxygen to drop.

Nitrogen/phosphorus can significantly affect aquatic life and long-term ecosystem health. High nitrogen and phosphorus loadings from wet weather discharges and overflows and non-point-source discharges may contribute to the production of harmful algal blooms, adversely affect fish habitats, contribute to fish kills, and produce hypoxic or "dead" zones. Furthermore, nutrient pollution can also affect public health via effects to drinking water sources from high levels of nitrates, possible formation of disinfection byproducts, and increased exposure to toxic microbes such as cyanobacteria.

2.0 BACKGROUND ON REGULATORY AND HISTORICAL PRACTICES

2.1 Requirements of the Clean Water Act

Discharges from separate and combined sewers are point-source discharges and, therefore, require a National Pollutant Discharge Elimination System (NPDES) permit to authorize the discharge. Combined sewer overflows are specifically addressed in section 402(q) of the act and are authorized through NPDES permits. In the 2004 *Report to Congress on Impacts and Control of Combined Sewer Overflows and Sanitary Sewer Overflows*, U.S. EPA explained that "SSOs that reach waters of the United States are point source discharges, and, like other point source discharges from municipal SSSs, are prohibited unless authorized by an National Pollutant Discharge Elimination System (NPDES) permit". Generally, U.S. EPA has not allowed sanitary sewer overflows (SSOs) to be permitted and, therefore, they are typically handled through federal or state enforcement orders.

2.2 Past Regulatory Practices

U.S. EPA issued two draft policies, one in 2003 and one in 2005, addressing treatment of peak flows at WRRFs. The 2003 draft policy allowed peak flows to be blended and discharged with certain constraints. The 2005 draft policy proposed that the 40 *CFR* 122.41(m) bypass regulation be applied to peak wet weather diversions at publicly owned treatment facilities serving separate sanitary sewer conveyance systems that are recombined with flow from the secondary treatment units.

In some areas of the country, U.S. EPA has been implementing the 2005 draft policy. This practice led to litigation and was eventually ruled on by the 8th Circuit Court of Appeals. For further information, the reader is directed to the 8th Circuit Court of Appeals March 2013 ruling (711 F.3d 844 [8th Cir. 2013]), which vacated U.S. EPA's current approach for addressing blending and remanded the matter back to U.S. EPA for further consideration. Based on the current situation, it is difficult to speculate on future regulatory trends.

3.0 NATIONAL POLLUTANT DISCHARGE ELIMINATION SYSTEM PERMIT CONSIDERATIONS

3.1 Technology-Based Limits and Water Quality-Based Limits

In section 301 of CWA, Congress directed U.S. EPA to establish technology-based limits for point-source discharges. In the case of industrial point sources, these standards are often referred to as "effluent guidelines". For publicly owned treatment works (POTWs), Congress required U.S. EPA to develop secondary treatment standards (performance standards, not specific technology requirements). Secondary treatment standards are described in 40 *CFR* Part 133. The standard provides for biochemical oxygen demand (BOD) and the total suspended solids (TSS) 30-day average of 30 mg/L and 7-day average of 45 mg/L. It also includes a percent removal standard for BOD and TSS of 85%, and pH must be controlled between 6 and 9 standard units.

Also in section 301 of CWA is the requirement that if the technology-based standard is not able to control the discharge to a level that will protect water quality standards, then the permitting authority must establish more stringent limits, that is, water quality-based limits (see CWA section 301[b][1][C]). In the regulations at 40 *CFR* 122.44(d), the permit writer is required to assess the discharge and determine if a reasonable potential exists to exceed water quality standards, either alone or in combination with other discharges.

Following is a brief summary of limits and requirements that may be applied to WRRFs. A more detailed discussion of these requirements can be found on U.S. EPA's Web site and in the following guidance manuals from U.S. EPA: http://cfpub.epa.gov/npdes/writermanual.cfm?program_id=45 and http://www.epa.gov/npdes/pubs/owm0264.pdf.

3.2 Concentration and Loading Limits

The terms, *flow* and *load*, are not synonymous. For the design and operation of some systems such as pumps and preliminary treatment, flow is the most commonly used parameter. However, for others including biological treatment systems and solids handling systems, loading is also critical. The permitting regulations require that limits be expressed in terms of mass whenever possible and provide for the use of concentration limits when the underlying standard is expressed in terms of concentration or the pollutant is more appropriately controlled as a concentration. The permitting authority may also include mass limits.

Wet weather conditions affect the temporal and spatial distribution of loadings at a WRRF. Influent concentrations will vary depending on the size, duration, and intensity of the storm event and if the collection system is separate or combined. In some separate and combined sewers that experience high wet weather flows, a first flush effect may occur distributing a high loading within the first few hours of an event and decreasing after the "flush" has passed. For other systems, the effect of wet weather is less pronounced, and no first flush is observed.

For systems that experience a pronounced "first flush effect", primary clarifiers and sludge processing systems may be overloaded if this phenomenon is not considered. For systems with high seasonal groundwater, a WRRF may experience difficulty in meeting loading limits as there must be a proportional reduction in some concentration limits to meet both concentration and loading limits in the NPDES permit. Ultimately, the reduction in concentration may be of the magnitude that additional treatment in the form of tertiary filters may be required to meet both concentration and loading limit parameters.

3.3 Seven-Day Average and 30-Day Average Limits

The secondary treatment standards provide limits in terms of 7-day and 30-day averages. In instances where these limits are not practical, other limitations may be applied. In the case of some toxic pollutants, such as metals, U.S. EPA, in the Technical Support Document for Water Quality-Based Toxics Control, recommended it is more appropriate to express the limits as daily limits and 30-day average limits instead of a 7-day average (see http://www.epa.gov/npdes/pubs/owm0264.pdf).

3.4 Bypass and Overflow Reporting

Reporting requirements vary greatly state to state and the definition of a *bypass* and an *overflow* (reportable overflow) often varies state to state. At the federal level, *bypass* is defined in the regulations at 40 *CFR* 122.41(m) as the intentional diversion of waste streams from any portion of a treatment facility. Paragraph (m) also includes the following requirements to be included in the NPDES permit for reporting the bypass:

> "(3) *Notice*—(i) *Anticipated bypass.* If the permittee knows in advance of the need for a bypass, it shall submit prior notice, if possible at least ten days before the date of the bypass.

(ii) *Unanticipated bypass.* The permittee shall submit notice of an unanticipated bypass as required in paragraph (l)(6) of this section (24-hour notice)."

Sanitary sewer overflows are not specifically mentioned in CWA or the regulations as CSOs are, and the reporting is generally believed to fall under the NPDES standard permit requirement to report noncompliance found at 40 *CFR* 122.41(l):

"(7) *Other noncompliance.* The permittee shall report all instances of noncompliance not reported under paragraphs (l) (4), (5), and (6) of this section, at the time monitoring reports are submitted. The reports shall contain the information listed in paragraph (l)(6) of this section."

The federal regulations at 40 *CFR* 122.41 include general reporting or information requirements that are to be included in NPDES permits.

4.0 CLIMATE CHANGE DESIGN CONSIDERATIONS

Climate change is occurring as demonstrated by research and empirical evidence. Studies have also indicated that changes are comparatively rapid in geological terms. Effects from this are already being experienced in different parts of the United States through increases in extreme events, heavy rains, and droughts. Overall, the long-term effect is expected to vary between mild to extreme.

Climate change is expected to affect wastewater utilities through changes in precipitation, less snowmelt, and higher temperatures, and possibly require higher flood plain protection as sea levels rise. Portions of the American West have already been experiencing reductions in rainfall, leading to unprecedented droughts. It is imperative that utilities perform risk analysis to determine how their utility may be affected. Presently, the exact nature and magnitude of effect is not fully known. Recognizing the high uncertainty associated with determining possible effects, it is important that future facilities' planning efforts be conducted to ensure that any plan developed be flexible and resilient to consider a wide range of precipitation, temperature, surface water levels, and groundwater levels that may occur. Presently, care should be taken to expending resources to mitigate future risks in which the nature and magnitude of effect is not well known. Additional information may be obtained from U.S. EPA's National Water Program Climate Strategy (U.S. EPA, 2012b) available at http://water.epa.gov/scitech/climatechange/upload/epa_2012_climate_water_strategy_section

IV_infrastructure_final.pdf. This Web site has a chapter on both wastewater and drinking water infrastructure.

5.0 DEFINITION OF FLOWS

The term, *flow*, can convey different meanings based on the situation being discussed. Certain aspects of the term, *flow*, pertain to the design of a collection and treatment system, whereas others are associated with operations at WRRFs. Commonly used terms for flow are found in Table 2.1.

6.0 INTRODUCTION TO SYSTEM WET WEATHER FLOW MANAGEMENT

6.1 Introduction

Based on regulations pertaining to the operation of WRRFs during wet weather conditions, this section provides general statements regarding a number of wet weather strategies to be discussed in this special publication.

6.2 Collection System Considerations

6.2.1 Collection System Type

Wet weather responses will vary depending on whether the system is separate, combined, partly combined, or separate with significant inflow and infiltration or high seasonal groundwater. The layout of the collection system may also be important. Typically, separate sanitary sewers are constructed in newer areas, while older "downtown" areas remain combined. These challenges require that a customized approach be developed for each situation to best access the wet weather responses under different seasonal and rainfall conditions and determine appropriate management strategies.

6.2.2 Flow Storage and Satellite Treatment

To manage wet weather flows, some utilities use flow storage in the collection system and WRRFs to maximize pollutant capture and reduce overflow activity. The benefits of storage include reduced size of conveyance, pumping, and treatment systems, and lesser reportable loading under the NPDES permit. Based on advances in supervisory control and data acquisition and instrumentation and control, there is an increasing degree of automation and control in operations to maximize storage and treatment and reduce

TABLE 2.1 Application of terms associated with flowrate.

Name	What it includes	How it is applied
Design flow	The average flowrate of effluent discharged through the final outfall for all days of the year.	NPDES permitting and load calculation for effluent discharge.
Average annual dry weather flow	The average flowrate of days without rain within the previous 72 hours. This flow is lower than the "design flowrate."	Typically used by operators in developing operational control plans.
Peak hourly flow (normal daily diurnal pattern)	This is typically associated with normal flow variation that occurs throughout the day and is seen most during normal dry weather conditions. Higher values typically occur during the day and lower values occur during the night.	Peak hourly flows are important as they help define loading requirements for biological systems that require aeration to meet NPDES permit requirements for oxygen-demanding substances such as BOD and ammonia-nitrogen.
Peak hourly flow (wet weather)	This is associated with peak flow over a 1-hour period. The 1-hour period is considered a unit of time that can be easily measured and recorded.	Design of headworks pumping stations, preliminary treatment, facility hydraulics, storage, and effluent discharge that moves flow through the facility during peak wet weather conditions.
Peak 2-hour flow (wet weather)	This is associated with peak flow over a 2-hour period.	The Texas Commission on Environmental Quality uses 2-hour peak flows as a permit condition. Some states have different criteria for flow conditions and must be consulted.
Peak instantaneous flow	This is typically associated with theoretical peak flow delivered to the doorstep of the WRRF. This may be higher than peak hourly flow, and is often not fully known because of flow dampening in large sewers.	Because of the uncertain nature of peak instantaneous flow and how it can be accurately determined, it is not typically used in the design of WRRFs.

(*continued*)

TABLE 2.1 Application of terms associated with flowrate (*continued*).

Name	What it includes	How it is applied
Sustained 24-hour flow	This is a term used to identify the highest sustained value of wet weather flow that may be treated by a WRRF over a 24-hour period.	Understanding the temporal nature of how wet weather flow affects the WRRF is important in helping to determine the best operational control strategy.
Maximum monthly flow	This is a term used to define the maximum flow that a WRRF may be required to treat over a peak month during a year.	Peak monthly flow conditions may be created through seasonal residence and high groundwater infiltration. The NPDES permit writers should consider this month in determining the "design flow".

reportable overflows. Alternatively, some communities have constructed satellite treatment systems to reduce overflows if new development would overload existing sewers and result in additional SSOs or CSOs when conveying flow to the WRRF.

6.2.3 Flow Storage, Satellite Treatment, and High-Rate Treatment

There are a number of options to help a utility reduce the effect of wet weather flows at a WRRF. These include system storage, satellite treatment, and high-rate treatment. Each must be evaluated for applicability, effectiveness, and cost on a case-by-case basis.

Flow storage is the most common approach taken in most communities. Storage can be provided in sewer systems, and in tanks or basins positioned at strategic points in the sewer system or treatment facility. Storage may also be inline or offline. Inline storage is common in sewer systems. Offline storage may be provided in tanks or basins. The size and type of storage system will depend on a number of factors, including size, available space on-site, frequency of use, and operation and maintenance requirements.

Satellite and high-rate treatment systems can also be used to reduce the wet weather effect on sewer systems and WRRFs in lieu of conveying wet weather flow to the WRRF. Again, the decision to use satellite or high-rate treatment must be done on a case-by-case basis. In most instances, utilities prefer to use storage systems if possible to avoid constructing another "discharging" facility into the receiving stream.

7.0 GREEN SOLUTIONS

There has been an increasing emphasis on the use of green infrastructure to address wet weather challenges in sanitary and combined sewer areas. Approaches to redirect storm flow away from sewer systems and reduce impermeable areas within the sewer shed through permeable pavement, bioswales, rain gardens, stream buffer strips, and other control measures have been shown to reduce SSOs and CSOs in problem areas and provide other benefits, including reducing the "heat island" effect in urban neighborhoods.

U.S. EPA has increasingly emphasized the importance of "green solutions" in addressing wet weather problems and promoting sustainability. As part of U.S. EPA's June 2012 guidance memorandum, *Integrated Municipal Stormwater and Wastewater Planning Approach Framework* (U.S. EPA, 2012a), commonly referred to as the *Integrated Planning Framework* or *IPF*, U.S. EPA encourages municipalities to incorporate green solutions.

8.0 REFERENCES

U.S. Environmental Protection Agency (2004) *Report to Congress on Impacts and Control of Combined Sewer Overflows and Sanitary Sewer Overflows;* (EPA-833/R-04-001); U.S. Environmental Protection Agency: Washington, D.C.

U.S. Environmental Protection Agency (2012a) *Integrated Municipal Stormwater and Wastewater Planning Approach Framework*. http://www.epa.gov/npdes/pubs/integrated_planning_framework.pdf (accessed May, 2014).

U.S. Environmental Protection Agency (2012b) National Water Program Climate Strategy. http://water.epa.gov/scitech/climatechange/upload/epa_2012_climate_water_strategy_sectionIV_infrastructure_final.pdf (accessed May, 2014).

3

Integrated System Approach for Design

Jim Scholl, Susan Moisio, and Nalin Sahni

Historically, urban wastewater and drainage systems have been developed to manage public health and to prevent flooding. Today, the focus is also on maintaining or restoring the water quality of the receiving stream. Historically, the water resource recovery facility (WRRF), the receiving stream, the collection system, and the surface water drainage system have been evaluated as separate systems with a linkage through an agreed-upon boundary condition. The current challenge is to move from consideration of these systems as separate to evaluation of these systems as an integrated whole. The integrated approach to design provides a process for integrated evaluation.

1.0 INTRODUCTION

This publication focuses on the WRRF; however, the best solution cannot be found without considering the collection system, the surface water drainage system, the land application site, and the receiving waterbody. An integrated planning approach focuses on the entire watershed.

The U.S. Environmental Protection Agency (U.S. EPA) has recently provided guidance on integrated planning. U.S. EPA published the *Integrated Municipal Stormwater and Wastewater Planning Approach Framework* in 2012. The framework provides guidance for utilities to address stormwater and wastewater in an integrated plan. The framework is voluntary and does not waive the Clean Water Act requirements, rather, it provides relief to utilities having numerous demands for their funds. The framework suggests six elements of an integrated plan. These elements are as follows:

- Develop a description of the water quality, human health, and regulatory issues to be addressed in the plan;
- Develop a description of existing wastewater and stormwater systems under consideration and summary information on the systems' current performance;
- Implement a process that opens and maintains channels of communication with relevant community stakeholders;
- Develop a process for identifying, evaluating, and selecting alternatives and proposing implementation schedules;
- Develop a process for measuring success; and
- Implement improvement to the plan.

The Water Environment Federation has also provided guidance for the integrated management of wet weather flows. Its guide, *Guide for Municipal Wet Weather Strategies* (WEF, 2013), states,

> "Integration of wet weather planning and design with ongoing system priorities for upgrading and rehabilitation is essential. Whenever possible, wet weather planning should be done holistically as a part of watershed planning when stormwater, habitat, receiving water conditions, floodplain, and other issues are considered and prioritized in context".

2.0 INTEGRATED SYSTEM PLANNING

Regulatory requirements for the WRRF, and the combined, sanitary, and storm system should be included in the integrated plan. Wet weather flows generated and conveyed by the collection system affect the WRRF. Simply maximizing flow to the WRRF without considering the entire system is not advisable. If the peak flows that can be generated in the collection system are sent to the WRRF without being mitigated, then the WRRF can be overwhelmed and treating this wider range of flows can be costly. Storage and conveyance systems (tanks, pipelines, pumps, gates, and weirs) with and without real-time controls (RTCs) can be evaluated and implemented as part of an integrated plan that addresses the collection system and treatment system risks. For example, the effects of maximizing flow to the WRRF can be evaluated in terms of the risk associated with the flows, considering the following issues:

- Damage to the treatment process (washout biomass, hydraulic flooding),
- Quality of effluent discharged not meeting the permit limits,
- Odor control, and
- Increasing the life cycle cost of treatment and pumping.

Risk factors in the collection system include the following issues:

- Sanitary sewer overflow (SSO) wastewater releases (e.g., flooded manhole or pumping station overflow),
- Combined sewer overflow and SSO discharges to waterbodies (constructed overflow locations), and
- Sedimentation issues in the collection system.

These effects can be evaluated and mitigated using integrated planning. The following elements should be considered:

- Storage in the collection system affects the peak flowrate to the WRRF during wet weather and prolongs flow to the facility after the storms have passed and flow is drained from storage units.
- Source control in the collection system can improve the existing capacity of the system by removing excess stormwater.
- Conveyance improvements, such as rehabilitation, cleaning, and upsizing of the pipes and pumping stations, can improve the capacity of the collection system.
- Real-time control of the collection system using flow control devices (pumps, gates, weirs) allows the flow regime (levels, flowrates, storage) to be changed in real time to better achieve multiple objectives. Satellite wet weather treatment facilities affect flows to the WRRF by removing these flows from the system at the location where the

satellite facility treats the wastewater and discharges it to the local stream. This introduces a potentially positive change to the volume and pollutant load previously experienced at the local stream.
- Passive control of the collection system through flow control devices (weirs, orifices) also allows the flow regime (levels, flowrates, storage) to be changed.

An integrated approach requires comprehensive knowledge of the system, communication between collection system operators and the WRRF, and sufficient hardware and programming to understand and evaluate the effects of operating an integrated system. A key component of this analysis requires the development and use of collection system models as presented in Chapter 9. Chapter 9 discusses details of how hydraulic and hydrologic modeling should be conducted.

3.0 INTEGRATED SYSTEM DESIGN COMPONENTS

The components considered in integrated system design are those elements that will help achieve the specific objectives that the overall system must address. Each system will have unique performance metrics, solutions, and approaches to control strategies. The performance metrics (i.e., flows, quality, storage, treatment targets, redundancy at peak wet weather flows, receiving water effects, floodplains, habitat, and watershed considerations) should be developed. Each watershed is unique, and the following site-specific variables with related requirements must be identified and established as project performance metrics to be accomplished by the facility and systems' design and operational solutions:

- Identify alternative solutions using the watershed approach (stormwater source control, inline actions, satellite or end-of-pipe systems, WRRF, and receiving waters).
- Control strategies should be integrated to accomplish key performance metrics. A key element will consider compiling life cycle costs to provide a decision process for comparing alternatives and documenting the selected mix of controls and processes to accomplish the targeted outcomes.
- Prepare the implementation plan for final design, construction, startup, and operation.
- Establish monitoring programs to document the results from implementation monitoring and ongoing evaluations.

- Evaluate performance compared to metrics to improve wet weather treatment performance on a scheduled basis.

4.0 INTEGRATED PLANNING EXAMPLE

An ideal integrated planning scenario would encompass the wastewater collection system, the WRRF, the surface water drainage system, and the receiving stream or land application site. The integrated planning process would include three objectives. Figure 3.1 shows an example of the components of the integrated planning process.

The completed process also includes construction of the selected plan and monitoring and evaluation of the constructed facilities. The monitoring and evaluation process is ongoing and should inform improvements to the facilities based on operational experiences.

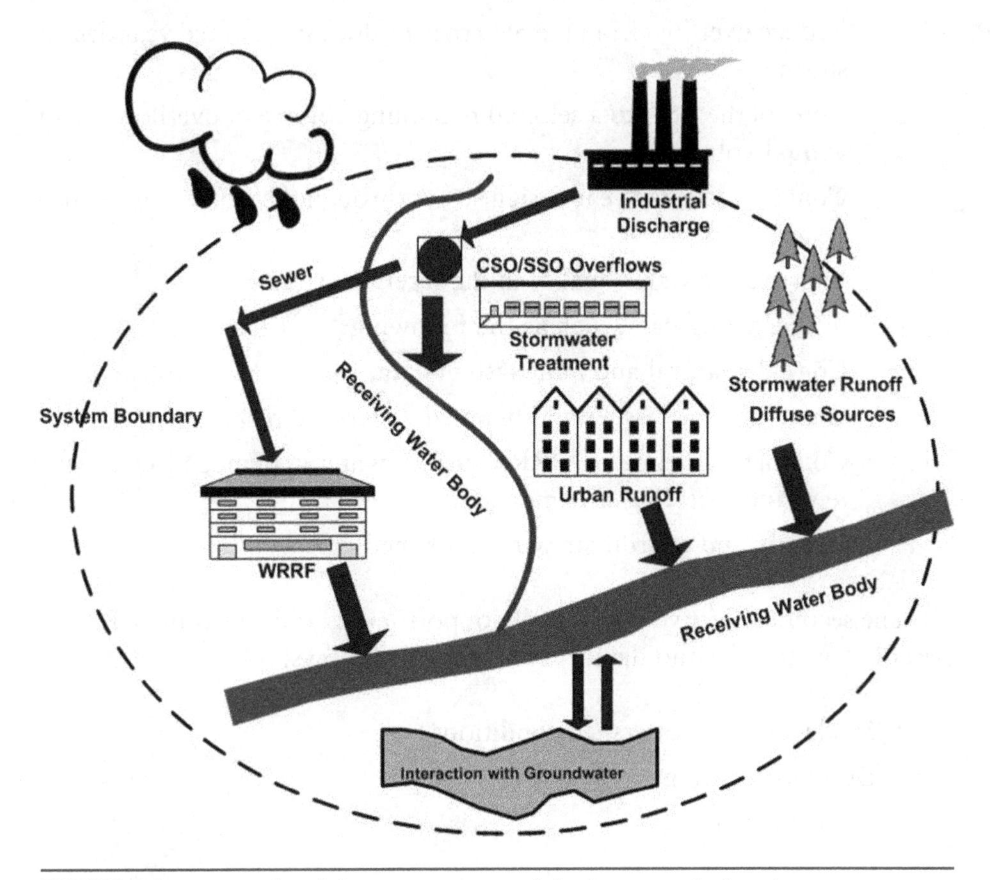

FIGURE 3.1 Example of the components of the integrated planning process.

An example of an ideal integrated planning scenario is as follows:

- Collection system description—combined and sanitary sewer system with storage tanks, RTC facilities to control flow, and storage/treatment facilities;
- Surface drainage system—piping and natural drainage system;
- Water resource recovery facility—primary and secondary treatment with an internal bypass after primary treatment and an outfall to the receiving stream. Bypassed flows receiving a minimum of screening for floatable control, primary sedimentation, and disinfection before discharge; and
- Receiving stream—recreational use stream.

The first objective is data compilation and inventory analysis. Examples of steps to take and data to collect are as follow:

- Define initial integrated plan goals and objects, for example:
  - Reduce overflow from the SSO to zero during a selected year design storm,
  - Control the CSO to a selected remaining volume of overflow on an annual volume basis,
  - Convey a selected year design storm through the storm conveyance system, and
  - Improve the water quality in the receiving stream.
- Collect existing data, such as the following:
  - Compile natural and built system data;
  - Compile policy issues and planned watershed projects;
  - Conduct modeling of WRRF, surface water system, collection system, and receiving stream; and
  - Identify and coordinate with watershed partners.

The second objective is to identify opportunities and constraints. Examples of steps to take and data to collect are as follows:

- Model current watershed conditions
- Develop a strategy:
  - Source control,

   - Evaluate and address the condition of existing infrastructure and systems,
   - Conveyance,
   - Storage,
   - Land application site—deep-well injection,
   - Increased WWRF wet weather capacity, and
   - Real-time control of the collection system.
- Identify opportunities, constraints, and the desired level of service:
   - Identify performance criteria such as no surcharging of pipes for selected rainfall events and
   - Identify opportunities for goals and objectives—examples include an opportunity to separate the surface water from a significant roadway, install infiltration trenches along streets in the combined sewer area, or build wet weather treatment facilities in the collection system to alleviate peak flows at the WRRF.
- Identify watershed constraints—examples include land constraints for the selection of sites for the wet weather treatment facility, soil constraints that could affect infiltration trenches, or subsurface geology that would prevent deep-well injection.
- Conduct an urban audit—the purpose of the urban audit is to provide information on the current status of the urban area and to influence the selection of sites for collection system elements such as stormwater retention ponds or other water features,
- Coordinate with watershed partners and revisit goals and objectives.

The third objective is to develop and evaluate alternatives. Examples of steps to take and data to collect are as follows:

- Identify watershed and subwatershed alternatives—conduct an evaluation and develop levels of service for alternatives: (an evaluation should include volumetric controls, effects on water quality, and effects on the WRRF process);
- Analyze cost estimates and affordability;
- Develop preliminary integrated plan with recommended projects:
   - Coordinate with watershed partners, revisit goals; and
   - Evaluate alternatives and refine to subwatershed level.

The fourth objective is to develop an integrated plan. Examples of steps to take and data to collect are as follows:

- Conduct capital improvement planning.
- Prioritize projects using a system such as triple bottom line. This system evaluates the proposed project or alternative using social, economic, and environmental factors.
- Engage community.

The steps after planning include design, construction, and evaluation of the selected plan.

5.0 REFERENCES

U.S. Environmental Protection Agency (2012) *Integrated Municipal Stormwater and Wastewater Planning Approach Framework*; U.S. Environmental Protection Agency: Washington, D.C.

Water Environment Federation (2013) *Guide for Municipal Wet Weather Strategies;* Water Environment Federation: Alexandria, Virginia.

6.0 SUGGESTED READING

Baughman, D.; Lodor, M.; Bush, C.; Chamblin, L.; Albertin, K.; Jean-Baptiste, S.; Moisio, S. (2013) *Sustainable Watershed Evaluation and Planning Process for Wet Weather Program Alternatives Development*; American Society of Civil Engineers: Reston, Virginia.

4

Integrated System Approach for Operations

James Drake, P.E., BCEE, and Jane McLamarrah, Ph.D., P.E.

1.0 INTRODUCTION

Although water resource recovery facilities (WRRFs) can be readily designed with the appropriate processes and capacity to treat target peak wet weather flows that are conveyed to the facility, a comprehensive and flexible integrated system operations approach is necessary to address the wide variability of operating goals and flowrates that occur during wet weather events. Operating decisions at the WRRF are made continuously throughout the event. Operating decision criteria should be formal so consistent operational protocol is followed by all operators during each event, but also flexible to meet the highly variable flowrate and operating conditions that might occur in the collection system.

A formal integrated operational approach is beneficial because it ensures that all staff understand the balance of utility operating priorities that must be achieved during every event, including the following:

- Protection of the collection system to avoid the surcharge of sewer into properties or streets,
- Compliance with effluent permit discharge criteria from upstream collection system (combined sewer overflow [CSO]/sanitary sewer overflow [SSO] and satellite treatment facilities) and WRRF outfalls, and
- Protection of WRRF processes and maximization of wet weather treatment at the WRRF without impairing the capability to treat dry weather flow after the event.

The utilitywide understanding of these goals should be incorporated to a set of wet weather standard operation procedures (SOPs) based on the conveyance capacity of the collection system, the availability and use of upstream (satellite) wet weather treatment/storage and control capabilities, and the design capacity of the WRRF. Ultimately, the goal of wet weather operations will be to maximize the use of all existing infrastructure without extending it to a failure point to minimize the discharge of untreated wet weather flow. In some instances, the utility operating the treatment facility may not be in control of upstream collection system facilities; thus, the ability to control influent WRRF flow and minimize upstream system discharges will not be within the utility's protocol.

2.0 PLANNING FOR INTEGRATED OPERATION

2.1 Effect of Collection System Maintenance

Maintenance of the collection system to maximize conveyance of flow to satellite treatment facilities and the WRRF is optimized and governed by capacity maintenance operations management (CMOM) programs that are implemented and enforced through various system discharge permits. A well-maintained collection system will help minimize upstream surcharging into properties and streets that is caused by temporary pipe constraints, but may not resolve inherent pipe capacity deficiencies or any collection system surcharging that might be caused by wet weather operations. Operators should be aware of the potential for CMOM programs to improve conveyance capacity, which will eventually increase flow to WRRFs during wet weather events.

2.2 Collection System Integration with Treatment

The use of real-time control (RTC) of collection system operations to control flow distribution and discharges from either CSO or SSO outfalls during wet weather events is becoming more prevalent. An integrated WRRF operations plan accounts for the flow control logic that might be applied during wet weather events. Accordingly, wet weather flow will be controlled by a hierarchy of treatment potential governed by system permit requirements. Real-time control can be used to prioritize flow conveyance options to the WRRF and then to storage options in the system (while avoiding upstream collection system surcharge and sewer service backups) to maximize the use of existing assets. Real-time control may also be used to direct flow to upstream storage or treatment facilities during the event, depending on the magnitude of the storm and the permit conditions applied to the receiving waterbodies. Operators of a WRRF should be aware of the possible range of RTC operations to maximize the amount of wet flow that is stored or treated in upstream facilities and to anticipate variable flowrates that are ultimately conveyed to the WRRF and the discharge effects on receiving streams. Holistic planning objectives for a wet weather system are discussed in *Guide for Municipal Wet Weather Strategies* (WEF, 2013).

Utilities also exercise care to balance possible effects in the collection system during severe storm events. Wet weather operations at the WRRF that could cause sewer service backups into properties, creating a potential public health hazard, are avoided, but these operations may result in potential receiving water quality exceedances.

Back-to-back storm conditions are also a challenge. On-site or upstream storage facilities may be full and the WRRF wet weather treatment should be maximized to empty storage in anticipation of the next storm. Screenings and solids from upstream treatment facilities may also be collected and conveyed down to the WRRF, requiring processing after the storm event.

2.3 Receiving Waterbody Considerations during Wet Weather Operations

Collection system control and WRRF operations must be guided by the discharge requirements for each of the outfalls in the collection system. There is a wide range of possible operational protocols that can be applied depending on the capacity of the collection system, the types of satellite wet weather treatment systems that are available, the use of blending at the WRRF, and the receiving streams (or groundwater) classifications. Upstream CSO/SSO outfalls may have different (and possibly more stringent) discharge limitations (e.g., fresh water or salt water). Typically, a WRRF provides the highest level of treatment to wet weather flow, but upstream satellite facilities may be capable of providing an equal or potentially higher level of treatment when blended flow scenarios at the WRRF are considered. If upstream waterbodies have less stringent discharge requirements during wet weather events, less sensitive uses, and/or the uses may be less impaired by upstream wet weather discharges, an integrated operations plan should be developed to ensure that wet weather flow is provided with the maximum treatment possible to meet receiving waterbody (or groundwater discharge) requirements.

2.4 Water Resource Recovery Facility

The integrated operations plan should emphasize WRRF preparedness in anticipation of the wet weather event (and any back-to-back storm conditions). A proper SOP will include pre-event checklists to inventory equipment that is available at the facility for treatment before the storm and flow "ramp-up" procedures to sequentially activate equipment and processes as the flow increases. In some instances, the capacity of the WRRF to accept additional flow simply means increasing pumping rates to put more flow in the tanks. In other instances, a WRRF may not be operating all tanks, pumps, or equipment during antecedent dry weather flow conditions. In these situations, to address wet weather treatment and to maximize the capture and treatment of wet weather flows, the WRRF operations plan will typically include activation of all redundant tankage and equipment to provide flow equalization throughout the facility. Dedicated processes and equipment may have special procedures that are required to be able to activate them during a storm. For example, chemical/process tanks may have to be filled before the wet weather event and operate with chemical injection, as a side stream, before the peak flows begin. Groundwater may have to be pumped out of tanks to ensure maximum capacity. Reducing facility recycle flowrates may also help to increase the WRRF capacity to treat influent wet weather flow.

To ensure that all equipment and tankage is available during the typical storm season, regular maintenance of process equipment should also follow prudent scheduling procedures to avoid equipment being down when storms are anticipated. In addition, good maintenance helps to minimize the failure of critical facilities during a storm event, when they are typically functioning at their peak design rates. Any failure of critical equipment at peak flow periods will likely result in water quality violations.

Wet weather preparedness is even more important for dedicated wet weather equipment and treatment trains that are not used regularly. Preparatory procedures include periodic inspection and testing of equipment (such as motors, gates, and screens) during interim periods and, potentially, more immediate checking just before the onset of the storm event. These preparatory procedures should be documented in SOPs to ensure a consistent state of readiness for storms.

In addition, the operations plan may incorporate process changes (such as a modification of the secondary biological process to step or contact feed aeration to store solids in the aeration tanks vs the secondary clarifiers or other internal process flow redirection strategies) to protect the WRRF from high flowrate (and dilute flow rate) effects. A plan must also be in place to revert back to normal processes (e.g., the use of step or contact feed aeration during a storm event requires a concurrent recovery strategy for moving back to plug flow following the event to avoid solids overloads to the secondary clarifiers during the transition back).

Typically, a wet weather event is characterized by a first flush with high solids that are conveyed to the WRRF. Comprehensive flow ramp-up procedures in the SOPs are important to capture and treat the first flush of the event while avoiding any process effects that could restrict treatment capacity during the latter portions of the wet weather event. Accordingly, the SOPs should consider operational limitations of all WRRF equipment under a range of operating conditions so that catastrophic equipment failures are not created during the event.

3.0 FACILITY OPERATIONAL CONSIDERATIONS DURING WET WEATHER EVENTS

3.1 Challenges of Precipitation Characterization during Wet Weather Operations

Wet weather operations are based on storm anticipation. Comprehensive storm tracking and characterization are necessary to determine the operational response to the incoming wet weather event. Depending on the size of the wet weather event, the approach pattern, rainfall coverage, and expected precipitation intensity and duration, different operational procedures may be activated to address the range of influent flow to the WRRF. Predictive computer modeling of both precipitation characteristics and collection system response is becoming more helpful to determine the type of flow and, potentially, pollutant load that may be conveyed to the WRRF. This information helps to inform the operators to staff the facility accordingly and to be prepared with adequate chemical storage and deliveries and equipment availability. Downstream notifications can also be generated by storm anticipation.

Storm tracking/predictive modeling should continue through the storm because of the variability of wet weather events even during the event itself. Monitoring of precipitation conditions during the event can help to manage the activation of equipment to avoid excessive shutdown and maintenance efforts after the storm.

3.2 Energy Management

Energy management at the WRRF during wet weather operations is an important consideration. Equipment and process activation plans to match peak flowrates can cause excessively high peak electrical demands, which may result in substantial peak electrical demand charges from the electric utility. Operators should be provided with training and monitoring equipment and flexible SOPs to manage electrical loadings at the facility during wet weather operations to avoid excessive peaks in electrical demand. Power monitoring and control systems may be needed to track power usage. More information on electric power use and management is provided in Chapter 7 of this publication.

3.3 Water Quality Effects of Regulatory/Permit Thresholds

An operational challenge at the WRRF during wet weather events can be the type of loads that are received at the facility. The first flush of pollutants (both from a quantity and quality point of view) during the wet weather event can overwhelm some portions of the process (i.e., grit facilities or primary clarifiers). This high turbidity of the influent that may carry through to the effluent can create difficulties for proper disinfection of the flow (or require much higher disinfection) to destroy the bacteria attached to the high turbidity solids. In contrast, flows in later periods of the storm event may be dilute.

Standard operating procedures should incorporate flow dampening (use of additional tankage or pipeline storage) and smooth operations to avoid intermittent or unexpected water quality violations. In addition, wet weather SOPs may dictate turning on additional equipment such as sludge pumps in anticipation of the storm to continually remove and convey solids away from potentially vulnerable processes, especially when storm events occur following particularly dry weather conditions.

3.4 Returning to Normal Operations and Minimizing Maintenance Requirements

A significant portion of wet weather operations at the WRRF is returning back to normal operations. Redundant equipment and tankage that were put online to handle the additional peak wet weather flow must be dewatered, cleaned, and return to standby status. This cleanup effort will take a defined amount of staff effort that is diverted from other ongoing operations and maintenance procedures.

In some instances, the equipment may be designed with automatic cleanup modes. This equipment should be monitored to ensure that the cleanup procedures were completed. Other equipment must be manually cleaned and SOPs should include systematic steps to ensure that the equipment is returned to a state of readiness for the next wet weather event. Special attention may need to be extended to chemical feed lines to ensure that they are properly cleaned out because these lines may be one of the most vulnerable to deposition and plugging in between storm events.

Accordingly, although activating all spare equipment in anticipation of the wet weather event to maximize the treatment potential at the WRRF would appear to be the most advantageous operating protocol, careful consideration should be given to the cost and effort required to deactivate equipment. Storm tracking and predictive modeling can help to balance the proactive readiness of the WRRF with the cost to deactivate equipment after the event. In addition, WRRF operations must anticipate the potential for back-to-back storm events (based on a review of weather forecast and predictive weather tools) to avoid excessive maintenance and cleanup of equipment that may be immediately reactivated for the next storm event. Activation and deactivation of facility processes and equipment should be reassessed periodically after events to see if SOPs should be adjusted to minimize poststorm deactivation and maintenance requirements.

3.5 Managing Stormwater Generated On-Site

When combined with high influent flows from the collection system, any pure stormwater flows to the WRRF from such sources as the site's return flow drains or storage basins compound peak flow volumes requiring treatment and further stress facility processes. Thus, it is important that facility operators prepare stormwater facilities along with treatment process ramp-up activities. Stormwater ramp-up activities include draining stormwater from storage basins or tanks and from secondary containment basins that might be required to contain overflows, leaks, or spills. Storage tanks that have accumulated groundwater should also be pumped out to provide capacity for on-site stormwater.

During the wet weather event itself, facility operators may be required to sample stormwater flows at each stormwater outfall. Routine sampling at stormwater outfalls is required under the National Pollutant Discharge Elimination System (NPDES) Industrial Multi-Sector Stormwater permits (U.S. EPA, 2009a). Water resource recovery facilities are considered "industrial" facilities under these permits. Depending on whether the U.S. Environmental Protection Agency (U.S. EPA) or the delegated state administers the state's industrial stormwater NPDES permits, specific stormwater permit requirements may vary from state to state, but may include routine daily visual inspection, quarterly stormwater discharge sampling, and annual benchmark monitoring (U.S. EPA, 2009b). Stormwater discharge pollutant limitations typically do not currently include a numeric pollutant limitation, but benchmark monitoring thresholds may be defined for specific parameters and it is possible that current benchmark monitoring values could become pollutant limitations in future general permit cycles.

The general permits also have requirements for implementation of stormwater best management practices (BMPs) under a site-specific stormwater pollution prevention plan. Operational stormwater BMPs are listed in Table 4.1 with a brief description of each BMP. The BMPs are focused on pollution prevention and reducing the amount of pollution discharged from the stormwater outfalls. Potential sources of stormwater pollution on the WRRF site include general site erosion and sediment; leaks, spills, or windborne splashing from open tanks and basins; spills and droppings from sludge handling and transport activities; dust from sludge incinerators or lime silos; chemical, fuel, and lubricant leaks or spills; and run-on from adjacent sites. The risk of contaminated stormwater from such sources may be particularly high during and immediately after a wet weather event when facility operators are focused on processing high flows through the facility. Proper control of WRRF site stormwater flows minimizes the potential for contaminated stormwater to exit the site and thereby adversely affects receiving water quality. Design considerations for stormwater management and reducing potential pollution sources are discussed in Section 3.3 in Chapter 5, titled "Site Development and Facility Arrangement."

TABLE 4.1 Stormwater management operational BMP examples.

BMP	Description
Good housekeeping measures	Clean grounds and secondary containment structures for trash and unnecessary debris. Keep material storage and handling areas clean and readily accessible. Prevent potential contamination of stormwater.
Daily routine observations	Visually monitor and report problems or conditions that potentially affect stormwater quality.
Quarterly routine inspections	Visually inspect all stormwater outfalls as required by the site's NPDES Multi-Sector General Permit.
Stormwater benchmark sampling and monitoring	Inspect, monitor, and sample all stormwater outfalls as required by the site's NPDES Multi-Sector General Permit.
Maintenance program for structural controls	Inspect, report, and correct problems with machinery or storage of sensitive items.
Storm inlet, drain, and outlet structure maintenance	Clean structural controls, such as grates, to remove floatables and debris to prevent and minimize stormwater contamination.
Erosion control measures	Inspect, report, and control erosion problems.
Spill prevention measures	Inspect for leaks or spills. Follow the site's spill prevention containment and countermeasure plan. Perform postleak/spill investigation and analysis to identify "lessons learned".
Spill response measures	Respond promptly to all leaks and spills. Block stormwater inlets in the area. Clean thoroughly without water unless area can be drained to process drains that return flow to the facility. Record all leak and spill events and report in accordance with the site's stormwater pollution prevention plan and applicable environmental regulations.
Employee training and education	Educate on all stormwater issues at least annually and for all new employees. Conduct spill response training exercises.

3.6 Postevent Analysis

The SOPs should include a postevent analysis phase. Using the facility logs and notes, precipitation and flow records, and water quality monitoring results, operations during the wet weather event should be evaluated to ensure that the SOPs were followed and to gauge their successful implementation. Understanding the type of storm and the flow generated by the system and conveyed to the WRRF helps to put each storm and postmonitoring into perspective (i.e., Can this magnitude storm be anticipated again and how often?, Should we enhance our SOP?, etc.). Where effluent quality was not maintained and/or the maximized flow was not treated, adjustments should be made to operating protocol to improve operations for the next storm. This self-assessment is critical in maintaining good operations, maximizing the use of existing equipment, protecting the WRRF, and maintaining compliance with the various regulatory permits. The postevent analysis can also identify capital improvement needs to ensure reliable and safe operations.

4.0 DEVELOPMENT OF STANDARD OPERATION PROCEDURES FOR WET WEATHER EVENTS

4.1 Wet Weather Configurations

Each WRRF will have a unique SOP that is prepared based on the functional process, type of offline and parallel equipment, and type of process that can be used to maximize treatment during the wet weather event. Successful SOPs are developed with the level of staff training in mind and incorporate existing procedures and practices. This section is intended to be a discussion of the general attributes that should be considered in the preparation of a WRRF wet weather SOP.

4.2 Purpose of Water Resource Recovery Facility Wet Weather Standard Operation Procedures

A WRRF wet weather SOP is a standard set of steps or actions that must be taken to operate a particular piece of equipment or process function. The purpose of the WRRF wet weather SOP is to maintain a consistency of approach between storm events and staff. A formal SOP ensures that variability in operating protocol is minimized and that quality and quantity of wet weather treatment is maximized using the existing assets. In addition, where failures of an SOP are noted in a postevent analysis, the SOP can be modified to avoid failures by others in subsequent storm events. The SOP also identifies priorities during wet weather operations. A number of objectives should be met during a wet weather event, but these objectives need to be balanced to avoid critical failures. Prioritizing the goals via an SOP helps formulate a consistent approach for the utility, which will optimize compliance with the NPDES permit and reduce untreated wet weather discharges.

The SOP also identifies the proper ramp-up decisions, including activation of equalization storage, if available, and the sequence of operations to start and maintain consistent wet weather treatment. Process components may require a pre-wet-weather startup so that they are operational before high flow conditions begin (i.e., filling tanks, etc.). Flow monitoring, instrumentation and control equipment, and flow sampling and testing intervals should also be included in SOPs. For example, flow volume paced sampling equipment may have to be adjusted to collect samples at different intervals to avoid overfilling sample containers during peak flow conditions, but then the equipment may have to be recalibrated to normal flow conditions to obtain the right sample volume. To cover conditions after and between rain events, the SOPs can also cover housekeeping and maintenance requirements after the storm has subsided. Emergency procedures and downstream notifications should also be incorporated to the SOP.

Above all, the SOP should be recognized within the utility as the "go-to" guide for protocols during the wet weather event; the SOPs should also be considered a "living" document, which shall be modified periodically as facilities are upgraded, equipment is replaced, new wet weather control strategies are adopted, climate change occurs, or discharge requirements change. The SOPs are also a key tool that can be used in annual training of operations and maintenance staff.

4.3 Standard Operation Procedure Parameters

Review of available wet weather facility SOPs shows that there are a wide variety of layouts using from one to several pages to describe the equipment, functional use, and operations and maintenance activities. The SOPs also range in detail from those covering a system or more sophisticated narratives that discuss unique equipment associated with the wet weather operations. This will all be dependent on whether there are dedicated systems for wet weather control or modified operations using existing equipment related to dry weather treatment. Along with the text, a focus plan/map and photographs of equipment, key features, and instrumentation control boxes may be invaluable to facilitate an understanding of the operational procedures.

General parameters typically included in the SOP are as follows:

- Unit process/functional process identification—this narrative is typically brief and covers the functional system overview and/or the unit process to be discussed. This may simply be a title and/or a listing of significant equipment associated with the functional process.
- Purpose and scope—this section identifies the purpose of the process and SOP and the general scope of the equipment/process operational parameters.
- Description of the system and the system components—an additional section may be provided in the SOP to provide a listing of all of the equipment associated with the discrete process or function, discuss the key equipment items and their significance to the process, and identify some of the ancillary processes or equipment that must be operated coincidentally with this specific process during wet weather events.
- Operational controls—wet weather processes at a WRRF may be controlled automatically by a supervisory control and date acquisition (SCADA) system or manually. The SOPs should include a detailed discussion of the sequence of operations and critical flow/load parameters that control the activation of equipment and/or flow diversion through the facility. A detailed discussion is beneficial not only for dedicated manual operations, but also for WRRF staff when SCADA controls may have to be overridden during severe storm events and/or in the event of an equipment/process mechanical failure.
- Operating procedures and responsibilities—this section should provide more detailed discussion and descriptions of the various wet weather operating procedures that must be implemented at the WRRF by personnel during the rain event. Detailed identification of trigger thresholds to initiate or perform each procedure (depending on discharge requirements, flow levels, equipment constrains, or capacity restrictions, etc., at the WRRF) should be incorporated to the SOPs. The procedures can be separated by pre-event, concurrent, and postevent categories depending on the function or equipment. Individual WRRF staff responsibilities to oversee, initiate, monitor, and shut down each piece of equipment or function should be detailed in the SOP.
- Preventive maintenance—preventive maintenance would identify the responsible party to perform the maintenance, intervals, and recommended procedures to ensure adequate startup and operational reliability of the process/equipment for storm events.

- Applicable regulations—the SOPs could include a section on the applicable discharge limitations for flows that receive treatment through the WRRF. This discussion could also include the requirements for wet weather discharges from the upstream portions of the collection system if these are relevant to operational discussions for the WRRF.
- Recordkeeping—completion of logs and records before, during, and after the wet weather event are important to document the conditions that were experienced and/or the situations that arose during the wet weather event. The WWTF rainfall event log will provide critical data for later, periodic self-assessment of the operational procedures to identify recurring problems and to make informed adjustments to the SOPs to improve operations, if necessary.
- Related documents and revisions—the SOPs should include notations of the date and general descriptions of the periodic adjustments that are made to the procedures after review and reassessment.

5.0 REFERENCES

U.S. Environmental Protection Agency (2009a) *Multi-Sector General Permit for Stormwater Discharges Associated with Industrial Activity.* http://www.epa.gov/npdes/pubs/msgp2008_finalpermit.pdf (accessed May 2013).

U.S. Environmental Protection Agency (2009b) *Industrial Stormwater Monitoring and Sampling Guide.* http://www.epa.gov/npdes/pubs/msgp_monitoring_guide.pdf (accessed May 2013).

Water Environment Federation (2013) *Guide for Municipal Wet Weather Strategies;* Water Environment Federation: Alexandria, Virginia.

6.0 SUGGESTED READINGS

U.S. Environmental Protection Agency (2009) *Developing Your Stormwater Pollution Prevention Plan: A Guide for Industrial Operators.* http://www.epa.gov/npdes/pubs/industrial_swppp_guide.pdf (accessed May 2013).

U.S. Environmental Protection Agency (2008) *Sample MSGP SWPPP Template (WORD).* http://cfpub.epa.gov/npdes/stormwater/msgp.cfm#msgp2008_swppp (accessed May 2013).

5

Site Selection and Facility Arrangement

Siddhartha Sengupta, P.E.; Valerie Fuchs, Ph.D.; Jared C. Hutchins, P.E., ENV SP; Jason Iken, P.E.; and Jane McLamarrah, Ph.D., P.E.

1.0 INTRODUCTION

To manage wet weather flows through the strategies outlined in Chapter 2 (e.g., flow storage and satellite treatment, flow storage and routing), utilities must adapt their existing infrastructure to convey and process the additional flows while meeting permit requirements. This process often starts with development of several alternatives, such as performing upgrades to existing collection systems and water resource recovery facilities (WRRFs) and using available land and/or acquiring new land and constructing new facilities. Evaluations are then conducted to determine the optimal solution(s) to carry forward into detailed design and construction stages.

Site selection and facility arrangement must both be considered when addressing wet weather treatment for WRRFs. Stakeholders involved in planning and design cannot address site selection while ignoring facility arrangements and, likewise, development of conceptual facility arrangements requires the consideration of potential sites. The basic tenets found in *Design of Municipal Wastewater Treatment Plants* (WEF et al., 2010) apply when addressing site selection criteria and facility arrangements. In particular, Section 4.2 of Chapter 3 covers present and future design requirements for WRRFs and Chapter 4 covers site selection and plant arrangement (WEF et al., 2010).

This chapter addresses key factors related to site selection and facility arrangements in the context of managing wet weather flows at existing and new WRRF sites. Several of these factors are included in Table 5.1 and discussed in further detail in Sections 2.0 and 3.0. Several factors are further illustrated using two case study scenarios in Section 4.0.

2.0 SITE SELECTION

The site-selection methodology for wet weather management will be project-specific and may include the following general elements:

- Identifying variables that influence WRRF siting;
- Factoring environmental, technical, institutional, and cost considerations;
- Developing and executing a WRRF siting evaluation plan;
- Identifying potential WRRF project components;
- Developing WRRF siting criteria;
- Identifying initial WRRF candidate results;
- Applying WRRF siting criteria; and
- Selecting a preferred WRRF site.

Variables influencing siting associated with wet weather management at WRRFs often involve regulatory drivers focused on enforcement of federal standards and evolving stormwater regulations; this topic is further detailed in Chapter 2, and can influence siting for WRRFs. Complex permit regulations involving wetlands and environmentally sensitive areas may be required when considering wet weather upgrades to WRRFs because the location of existing infrastructure, including interceptors and outfalls, is fixed and often located within heavy industrial areas and waterways. Refer

TABLE 5.1 Typical WRRF siting and arrangement considerations for management of wet weather.

Category	Issue	Siting and arrangement considerations
Environmental	Potential effects on waterways	Avoid by using existing outfall. May require WRRF outfall modifications, including incorporation of junction boxes and upsizing or "twinning" of an existing WRRF outfall, or incorporation of a dedicated outfall. May require utility relocation. Survey activities should be conducted.
		Minimize or mitigate effects by incorporating features to garner mitigation credits from regulators (e.g., establishment of aquatic habitats, use of conservation easements).
	Potential effects on "brownfield"	Avoid by using low-impact construction methods (tunneling).
		Avoid by building vertically on existing facilities.
	Conflict with adjacent land use	Minimize disruption to surrounding properties during construction and operations.
		Incorporate buffer.
Technical	Limited support utility infrastructure	May co-utilize existing WRRF infrastructure, such as chemical storage and feed systems, sludge management, administration facilities, and parking.
		Residuals management associated with auxiliary wet weather treatment may dictate on-site storage.
	Contractor operations	Geologic conditions will influence contractor excavation methods, and retrofit of existing WRRF sites may limit the amount of space for excavation layback, stockpiling, and staging.
		Existing WRRF infrastructure, such as unused lagoons, may provide temporary groundwater management.
		Space for queuing of haul trucks, need for temporary construction entrances, and effect on existing WRRF operation should be considered.
		Multiple contracts and overlapping areas may be required to execute complex upgrade projects while maintaining existing WRRF service and meeting regulatory deadlines.

(*continued*)

TABLE 5.1 Typical WRRF siting and arrangement considerations for management of wet weather (*continued*).

Category	Issue	Siting and arrangement considerations
Technical (*continued*)	Intermittent process requirements	Sodium hypochlorite solution strength degrades slowly over time, which influences bulk storage tank sizing.
		Chemical volumes must be sufficient for the next event, which influences bulk storage tank sizing.
		Access provisions for removal of grit should be provided.
		Polymer storage design should consider the shelf life of the stored polymer product. Dry products may require a low-humidity atmosphere, and liquid products tend to degrade over time.
Institutional	Regulations	Evolving stormwater regulations may lead to overly conservative sizing.
		Lack of consensus on performance requirements may lead to overly conservative sizing.
		Consent decrees may dictate siting, sizing, and construction schedules.
	Land acquisition	May be required for RTC or satellite treatment facilities within a watershed, and to expand the existing WRRF site.
		Consider consolidation vs expansion of facilities.
		Minimize effects by collaborating with stakeholders for integration with public amenities (e.g., parks, trails, public education).
Cost	Funding for imposed improvements	Additional funding may be required for land acquisition.
		When costs of common elements can be shared between wet weather facilities and WRRFs, overall facility costs can be lowered, which may affect site selection.

to Section 4.1 for an illustration of regulatory and permit-related factors and their influence on siting.

As part of a siting evaluation, key stakeholders will often develop wet weather siting alternatives focused on the use of available space at the existing WRRF site to avoid additional land acquisition. These siting alternatives typically involve consideration of emerging technologies with reduced footprints to provide wet weather treatment upgrades while minimizing effects to the site. Sections 4.1 and 4.2 illustrate examples in which limited buildable space is addressed with small footprint technologies to upgrade existing WRRF infrastructure for auxiliary wet weather treatment. Chapters 7 and 10 through 16 can be consulted for typical design guidelines for unit processes common in wet weather flow management strategies.

Some specific technology considerations influencing site area requirements are as follows:

- Upstream equalization can be an effective approach to managing flow variations and reducing effects on downstream processes and can reduce equipment sizes and space requirements. Risks associated with system storage such as odor potential and land acquisition should also be considered (see Chapter 10).
- Designers should seek opportunities to use existing WRRF amenities for auxiliary wet weather treatment facilities. For example, it may be possible to use existing facilities for waste materials and residuals handling, chemical storage and feed, vehicle storage and parking, and public access and site safety to support the addition of auxiliary wet weather treatment at a WRRF, thus reducing site area requirements.
- Designers should provide sufficient area for roadways, sidewalks, and equipment staging for periodic solids and grit removal. Intermittent and widely varying flow streams and facility draining, which are common wet weather operations, tend to settle solids and grit in channels, basins, and transition areas despite thorough design efforts to avoid them. Large mobile cranes with clamshells, vacuum trucks, and dump trucks should be provided access near the channels and basins to remove grit. Alternatively, a ramp into the channels and basins can be provided to allow for grit removal with a front loader and dump truck.
- Sludge management for auxiliary wet weather treatment processes may require on-site storage with sufficient volume to handle long weekends or holidays (refer to Chapter 16). Additionally, remote intermittent-use wet weather facilities located within the collection system may require on-site storage for solids management during periods when the interceptor is surcharged and unavailable to convey solids to the downstream WRRF.

3.0 FACILITY ARRANGEMENT

3.1 Layout Types

The three basic layout types include linear, campus, and compact. Linear and campus WRRF layouts are typically easier to expand and/or retrofit for additional wet weather treatment capacity and typically have greater site area and funding requirements. Existing WRRF site space constraints may require consideration of compact auxiliary wet weather treatment facilities (see Section 4.2).

3.2 Arrangement of Treatment Processes

Common WRRF unit processes are described in detail in subsequent chapters. Designers should be mindful of existing WRRF operational requirements, vendor equipment options, maintenance access requirements, anticipated traffic routes, and areas reserved for future expansion. Designers should pay particular attention to the following arrangement considerations for management of wet weather:

- Water resource recovery facility upgrades, or satellite storage/treatment options, must consider the upstream hydraulic grade line and facility freeboard requirements noted in Chapter 6, while minimizing or avoiding effects on the upstream system. Solids deposition and basement flooding are two such issues that are addressed in Chapters 9 and 10.
- Upgrades to an existing WRRF may require that unit processes remain in operation with limited outages. This can influence layouts by requiring staged construction and connections that can be made with minimal disruption to WRRF operations. Locating key equipment out of flood-prone areas is good practice to mitigate effects from wet weather disasters; refer to Section 3.3, "Site Drainage," for specific guidance.
- Facility geometry may be dictated by existing WRRF infrastructure. Common wall construction of basins with influent and effluent channels may present particular challenges when attempting to create new flow paths; refer to Chapter 10, "Onsite Storage/Flow Equalization", for further discussion.
- Certain maintenance needs are magnified at intermittent-use wet weather facilities, including dewatering and flushing, management of chemical inventories, and grit handling/disposal. Designers should work with operations and maintenance (O&M) staff and vendors to anticipate areas requiring access and design the facility to facilitate

access. For example, wet weather facilities often demand robust screening and pumping equipment, which may require periodic servicing resulting from storm debris present in combined sewer systems. Consider space requirements for flushing and/or odor control systems to manage odors.

3.3 Site Drainage

Traditional emphasis on-site drainage for WRRFs has focused on stormwater quantity. This emphasis has led to the need to relocate storm drains and storm sewers during facility expansions and when wet weather facilities are added. It is important to consider ultimate facility footprints when designing on-site stormwater drainage facilities. Managing runoff quantity over the life of the WRRF is most flexible if as much stormwater as possible is allowed to infiltrate on-site.

Also critical to stormwater management at facility sites is the need to consider where and how stormwater will be discharged to the receiving stream. Some facility sites require flood protection berms to protect the site from stream flooding. These protections are designed for the predicted 100-year flood event. With the potential weather effects associated with climate change or with increases in the amount or nature of upstream watershed development, recent or future predictions of 100-year elevations could increase and require the berms to be raised.

The existence of protective berms or the location of a particular stormwater outfall may prevent gravity discharge to the receiving stream, thus requiring the stormwater to be pumped to the receiving stream. Under wet weather conditions, the stormwater pump station is another "process unit" that must be maintained and kept in working order by facility operators under adverse conditions. Backup power or pumps may be needed to prevent the site from flooding as a result of a failed stormwater pumping station.

Just as redundancy is important for critical process units, redundancy is important for stormwater facilities. This redundancy could take the form of adding resiliency to buildings or other site facilities, such as adding flood protection doors to ground-level building entrances or elevating access/egress points. Facility design should always ensure safe access to and from the site during flood events for facility personnel and emergency equipment. Appropriately designed flood walls can protect headworks, below-ground access locations, and open basins from flooding resulting from large storms.

In addition to the on-site stormwater drainage issues, stormwater runoff from upstream or adjacent properties should be considered. The quantity of the runoff is an important consideration in sizing stormwater facilities, but the routing of runoff should also be critically evaluated. As noted in Section 3.5 of Chapter 4, WRRFs are considered "industrial" facilities per federal and state regulations under National Pollutant Discharge Elimination System (NPDES) Industrial Multi-Sector Stormwater permits. In most delegated states, these permits are issued as general permits and the specific requirements may vary from state to state (U.S. EPA, 2009). However, all of the general permits are required to be at least as stringent as the federal multisector permit guidelines, which require quarterly and annual sampling of runoff from the industrial site. Water resource recovery facilities that commingle on-site stormwater with off-site runoff are sampling stormwater from other parts of the watershed in addition to the facility site contribution. If the state's general permit contains numeric stormwater quality limitations or even benchmark sampling reporting, the WRRF sample results could exceed those limitations or benchmark measures.

Consequently, it is recommended to segregate stormwater flowing onto the WRRF site from upstream or adjacent properties from stormwater runoff from the WRRF. Further, it is recommended that each stormwater outfall be installed in a manner that facilitates closing the outfall should an emergency situation or a spill event occur that has the potential to contaminate stormwater running off the facility site.

Water resource recovery facilities largely control the amount of pollutants in stormwater discharges from the facility site by implementing pollution prevention measures. These pollution prevention measures are in the form of stormwater best management practices (BMPs) that are designed to properly manage stormwater and to improve stormwater quality exiting the site. Table 5.2 lists example BMPs that should be considered during facility designs. Operational stormwater BMPs are discussed in Chapter 4.

Infiltration basins are typically used to reduce the amount of stormwater runoff exiting the site. During wet weather events, these facilities can capture "first flush" pollutants by controlling initial site runoff and discharging excess stormwater through the overflow pipe. Vegetated drainage swales perform a similar function in slowing stormwater runoff and allowing some infiltration of the stormwater.

Low impact development (LID) and *green infrastructure* are relatively broad terms, but, in this context, are used to refer to such stormwater pollution prevention measures as porous pavement (typically in visitor or parking areas with less potential for contamination from wastewater, chemicals, and sludge), bioretention systems, rain gardens, water barrels, and curb bump-outs to mimic natural hydrology, promote infiltration, and stop pollutants from running off-site. Green infrastructure relates to stormwater as a network of open spaces that naturally reduce flood risk and improve water

TABLE 5.2 Examples of stormwater design BMP.

BMP	Description	Potential benefits
Infiltration basin (also known as *recharge basin*)	A shallow artificial pond designed to infiltrate stormwater through permeable soils to the groundwater aquifer and includes an overflow pipe or other outlet for wet weather events.	Manages stormwater runoff. Prevents flooding. Prevents downstream erosion. Improves water quality.
Vegetated drainage swales	Broad, shallow channels with dense vegetation on side slopes and the bottom.	Traps particulate pollutants (suspended solids and trace metals). Promotes infiltration. Reduces flow velocity.
Low impact development	Bio-infiltration systems and site planning to minimize soil compaction and preserve vegetation; designed to mimic or restore natural hydrology.	Infiltrate rainfall on-site. Filter particulates. Potential removal of oils and metals through adsorption.
Green infrastructure	The use of natural systems, or engineered systems that mimic natural systems to treat polluted runoff. These nature-based systems are networked together to reduce flood risk and improve water quality.	Integrates various land uses and ecosystem services. Supports long-term sustainability. Positively influences human health.
Visible, accessible stormwater outfalls	The locations where stormwater exits the facility, including pipes, ditches, swales, and other structures that transport stormwater and that are designed to be visible and easily accessible.	Provides safe access for visual inspection of stormwater discharges. Provides safe access for stormwater sampling.

quality (Wise et al., 2010). As noted in Chapter 2, the U.S. Environmental Protection Agency (U.S. EPA) strongly encourages the use of green infrastructure approaches to manage wet weather. Since 2007, U.S. EPA's Office of Water has released four policy memoranda supporting the integration of green infrastructure to NPDES permits and combined sewer overflow (CSO) remedies (U.S. EPA, 2007a, 2007b, 2011a, 2011b). The incorporation of green infrastructure to new facility or facility expansion designs can provide an opportunity to include public demonstration facilities for innovative green infrastructure facilities for public education. For example, Clean Water Services in Portland, Oregon, is developing a wetland/filtration/pond system as a polishing mechanism, stormwater management technology, and park/public demonstration site.

4.0 CASE STUDIES

Example site selections and facility arrangements used for wet weather management at WRRFs are discussed below.

4.1 Adding Wet Weather Capacity at an Existing Water Resource Recovery Facility

This scenario involves the addition of 878 ML/d (232 mgd) of peak wet weather auxiliary treatment and 227 ML (60 mil. gal) of equalization storage at the Bay View Water Reclamation Facility in Toledo, Ohio. The consent decree required Toledo to construct wet weather storage and treatment facilities to treat up to 738 ML/d (195 mgd) through advanced secondary treatment, to store 227 ML (60 mil. gal) of additional flow for later advanced secondary treatment, and to treat any remaining excess flow up to 1514 ML/d (400 mgd) through high-rate, chemically enhanced primary treatment and disinfection.

Nine equalization basin alternatives were evaluated, which included a combination of basin configurations, depths, and locations. Monetary and nonmonetary factors were considered. Water resource recovery facility staff, affected community representatives, and project team members selected importance factors and weighting factors, which were applied to the nonmonetary factor scores and combined equivalent rating scores, respectively, to develop total alternative rankings.

Two equalization basin locations were evaluated: Harrison's Marina and Retiree's Golf Course (see Figure 5.1). The marina site would require significant dewatering and environmental permitting resulting from its proximity to the shore; relative isolation from Summit Street makes aesthetics and odors less of a concern. The golf course site location would result in the conversion of three of 12 holes for use in constructing the equalization basins, and would require the city to provide upgrades to the golf course as compensation. The golf course location would be in a more visible location along Summit Street, thus requiring aesthetic exterior finishes; the site would also be easier to construct upon and would require less environmental permitting. Both sites required deep friction piles for construction, with the

FIGURE 5.1 Bay View Water Reclamation Facility equalization basin alternatives.

marina site presenting bearing and uplift concerns and the golf course site presenting uplift concerns as a result of rising groundwater.

Two equalization basin configurations were considered: one 227-ML (60-mil. gal) multicell rectangular reinforced concrete tank and six 38-ML (10-mil. gal) prestressed concrete circular tanks. Dedicated first-flush compartments for the equalization basins would be covered and the remaining equalization volume would be either covered or uncovered. A system of flushing gates, water cannons, and nozzles would provide for flushing of the basins. The rectangular tanks could take advantage of common wall construction, whereas the circular tanks would require interconnecting pipes and tall profiles resulting from the dome heights required for tank covers. The rectangular basins with covers on only the first-flush compartments were featured in the selected alternative.

The selected alternative, "Option 5—Golf Course above Ground Mostly Uncovered", was carried forward into the facility evaluation process and a 95-ML (25-mil. gal) first phase facility was ultimately constructed in 2004 along with grit facilities, ballasted flocculation and disinfection, and effluent pumping and piping upgrades. To meet the aggressive consent decree schedule requirements and high construction staffing requirements and to minimize

WRRF outages, construction was divided into four separate contracts (see Figure 5.2). A master site plan and construction coordination specification section was shared among the four construction contracts to define construction sequencing and responsibilities at the common boundaries of the construction contracts. The last contractor on the site did the final paving of common access roads and landscaping of new and disturbed project areas.

4.2 Comparison of Conventional and High-Rate Treatment Technologies at a New Site

This scenario considers two approaches to providing 379 ML/d (100 mgd) of wet weather treatment for CSO control as part of the King County, Washington, CSO Pilot Program. King County's Regional Wastewater Services Plan (2005) called for development of satellite CSO treatment facilities. The 5-year CSO program update from 2005 recommended that King County continue to monitor high-rate treatment for potential consideration for

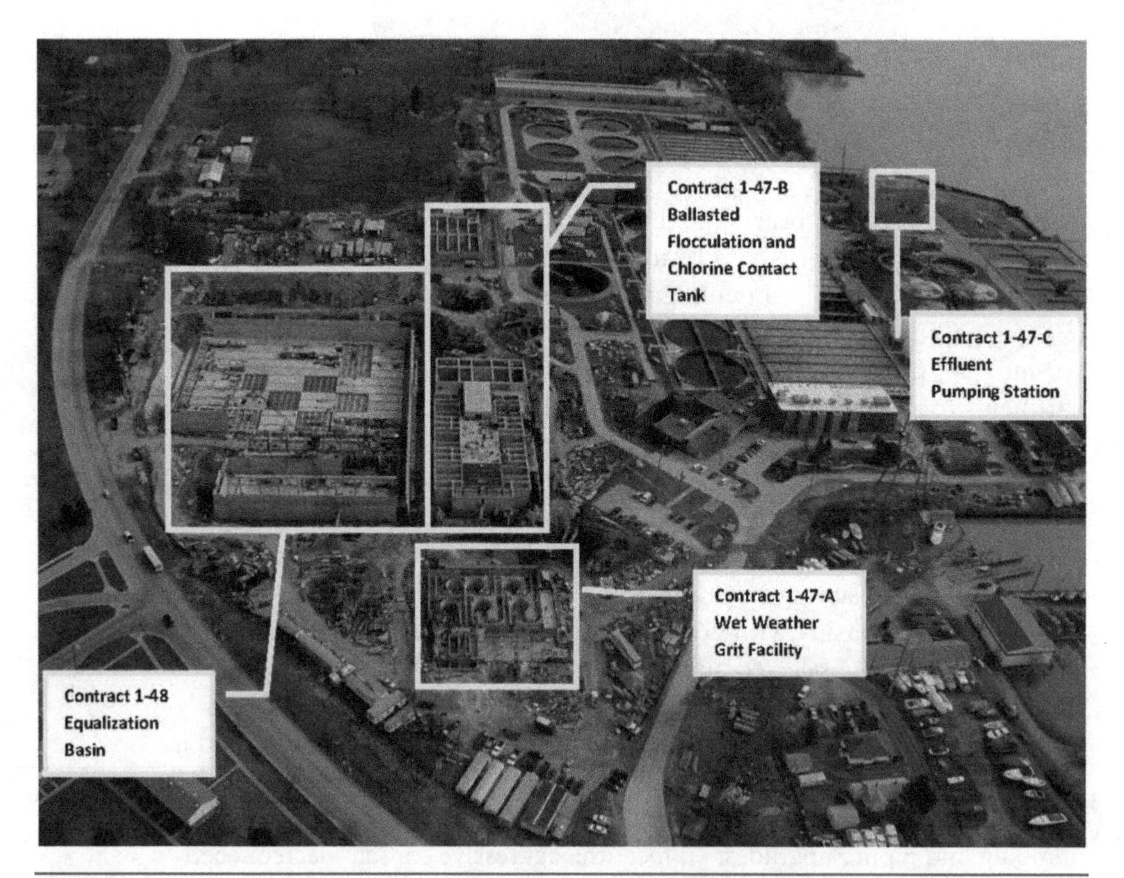

FIGURE 5.2 Bay View Water Reclamation Facility equalization basin alternatives.

CSOs. Similarly, the 2006 CSO Control Program Review recommended that pilot tests "be conducted on promising new CSO treatment technologies". A high-rate treatment (HRT) alternative was chosen to illustrate the differences in site requirements compared to conventional treatment technologies.

As part of this evaluation, several sites along the Duwamish waterway where existing CSO outfalls are located were considered for a hypothetical conversion to CSO treatment facilities meeting Department of Ecology (2006) CSO treatment guidelines (50% total suspended solids reduction, .03 mg/L·h settleable solids). Two non-site specific figures were generated to illustrate the overall footprint requirement for conventional vs HRT for a limited use facility. Footprint requirements for all process units, not just the liquids/solids separation unit, were considered and appropriate facility arrangements were developed. Adequate spaces were provided for necessary ancillaries, including odor control, pumping stations, standby power, operations, and site buffers.

- Figure 5.3 outlines a conventional WRRF that includes a peak discharge rate of 1.5 $L/m^2 \cdot s$ (2.2 gpm/sq ft) loading rate. The initial fill cycles of the sedimentation basins would serve as on-site equalization basins before discharging to downstream processes, allowing for 19 ML (5 mil. gal) of storage serving effectively as retention storage basins. Conservative design standards were used in this exercise for estimating chlorine contact time and dosing requirements for sizing. Standard vehicle turning radiuses were used from the Washington Department of Transportation for chemical deliveries, screenings removal, and facility maintenance.
- Figure 5.4 outlines the facility layout for a compact HRT process based on a peak discharge rate of 41 $L/m^2 \cdot s$ (60 gpm/sq ft), loading rate and related unit process tankage for coagulation and maturation zones, clarification zones, and relevant influent and effluent channels and pertinent sludge pumping and handling located inside the dedicated HRT footprint. In this scenario, two process trains would be used to split flow evenly to provide some level of redundancy for a limited use facility (CDM Smith, 2010). A storage basin is located upstream, providing an equalization volume of approximately 8.3 ML (2.2 mil. gal) serving as a retention storage basin before facility startup. Coagulant and polymer system designs were based on piloting experiences at King County and correlated with other operational facilities using 40 and 1.5 mg/L, respectively. Standard vehicle turning radiuses were used from the Washington Department of Transportation for chemical deliveries, screenings removal, and facility maintenance.

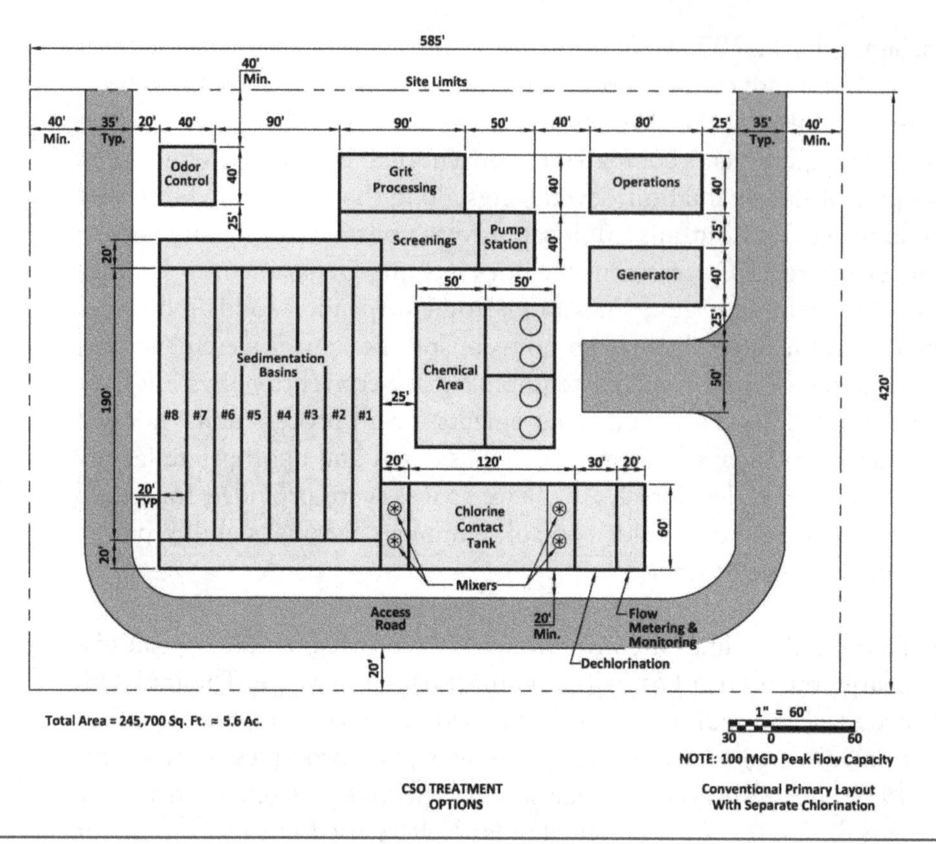

FIGURE 5.3 Conventional WRRF layout example.

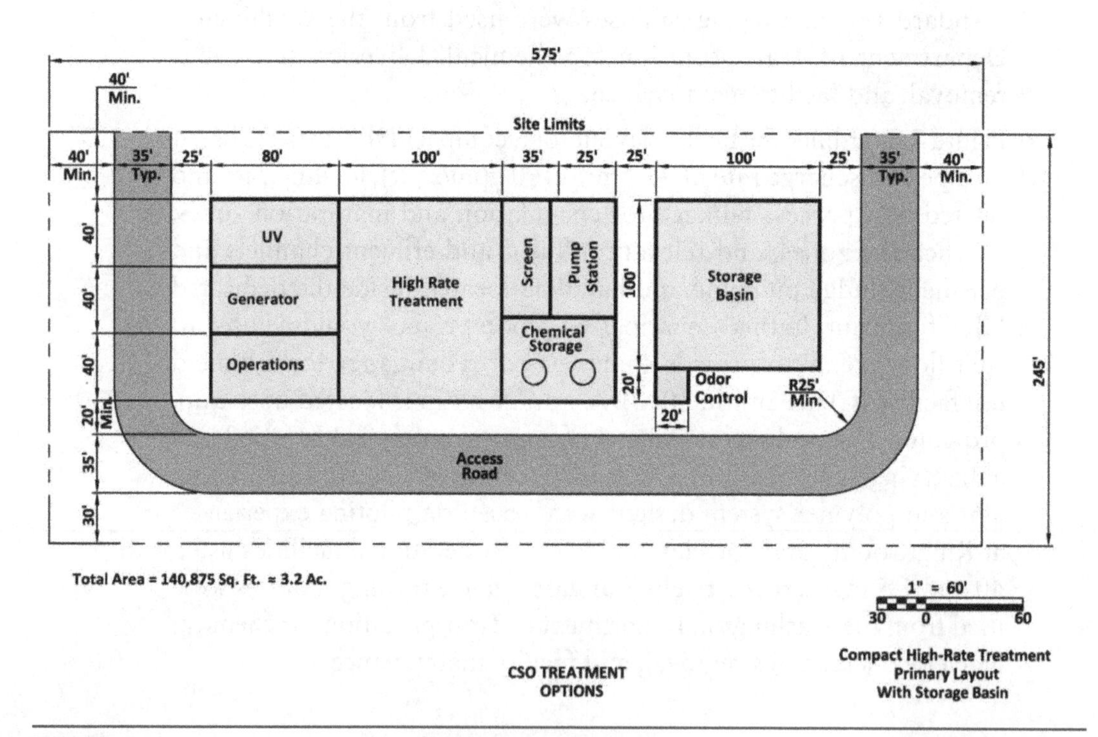

FIGURE 5.4 High-rate treatment WRRF layout example.

5.0 REFERENCES

CDM Smith (2010) Combined Sewer Overflow Treatment Systems Evaluation and Testing Piloting Report; CDM Smith: Seattle, Washington.

Department of Ecology (2006) Orange Book Standards, Criteria for Sewage Works Design: Chapter C3, Combined Sewer Overflows.

King County Wastewater Treatment Division (2005) King County's Regional Wastewater Services Plan; King County Government: Seattle, Washington.

King County Wastewater Treatment Division (2006) CSO Control Program Review; King County Government: Seattle, Washington.

U.S. Environmental Protection Agency (1999) *Combined Sewer Overflow O&M Fact Sheet Proper Operation and Maintenance*; EPA-832/F-99-039; U.S. Environmental Protection Agency: Washington, D.C.

U.S. Environmental Protection Agency (2011b) *Protecting Water Quality With Green Infrastructure in EPA Water Permitting and Enforcement Programs.* http://water.epa.gov/infrastructure/greeninfrastructure/upload/gi_memo_protectingwaterquality.pdf (accessed May 2013).

U.S. Environmental Protection Agency (2009) *Developing Your Stormwater Pollution Prevention Plan: A Guide for Industrial Operators;* EPA-833/B-09-002; U.S. Environmental Protection Agency: Washington, D.C.

U.S. Environmental Protection Agency (2007a) *Use of Green Infrastructure in NPDES Permits and Enforcement.* http://water.epa.gov/infrastructure/greeninfrastructure/upload/gi_memo_enforce.pdf (accessed May 2013).

U.S. Environmental Protection Agency (2007b) *Using Green Infrastructure to Protect Water Quality in Stormwater, CSO, Nonpoint Source and other Water Programs.* http://water.epa.gov/infrastructure/greeninfrastructure/upload/greeninfrastructure_h2oprograms_07.pdf (accessed May 2013).

Water Environment Federation; American Society of Civil Engineers; Environmental & Water Resources Institute (2010) *Design of Municipal Wastewater Treatment Plants,* 5th ed.; WEF Manual of Practice No. 8; ASCE Manuals and Reports on Engineering Practice No. 76; Water Environment Federation: Alexandria, Virginia.

Wise, S.; Braden, J.; Ghalayini, D.; Grant, J.; Kloss, C.; MacMullan, E.; Morse, S.; Montalto, F.; Nees, D.; Nowak, D.; Peck, S.; Shaikh, S.; Yu, C. (2010) *Integrating Valuation Methods to Recognize Green Infrastructure's Multiple Benefits;* Center for Neighborhood Technology: Chicago, Illinois.

6.0 SUGGESTED READINGS

Crow, J.; Smyth, J.; Bucher, B.; Sukapanotharam, P. (2012) Treating CSOs Using Chemically Enhanced Primary Treatment With and Without Lamella Plates—How Well Does It Work? *Proceedings of the 85th Annual Water Environment Federation Technical Exhibition and Conference* [CD-ROM]; New Orleans, Louisiana, Sept 29–Oct 3; Water Environment Federation: Alexandria, Virginia.

Fitzpatrick, J.; Bradley, P.; Duchene, C.; Gellner, J.; O'Bryan, C.; Ott, D.; Sandino, J.; Tabor, C.; Tarallo, S. (2012) Preparing for a Rainy Day—Overview of Treatment Technology Options for Wet-Weather Flow Management. *Proceedings of the 85th Annual Water Environment Federation Technical Exhibition and Conference* [CD–ROM]; New Orleans, Louisiana, Sept 29–Oct 3; Water Environment Federation: Alexandria, Virginia.

U.S. Environmental Protection Agency (2007) *Wastewater Management Fact Sheet In-Plant Wet Weather Peak Flow Management;* EPA-832/F-07-016; U.S. Environmental Protection Agency: Washington, D.C.

U.S. Environmental Protection Agency (2011) *Achieving Water Quality Through Integrated Municipal Stormwater and Wastewater Plans.* http://water.epa.gov/infrastructure/greeninfrastructure/upload/memointegrated-municipalplans.pdf (accessed May 2013).

U.S. Environmental Protection Agency (2013) *Emerging Technologies for Wastewater Treatment and In-Plant Wet Weather Management;* EPA-832/R-12-011; U.S. Environmental Protection Agency: Washington, D.C.

Water Environment Federation (2013) *Guide for Municipal Wet Weather Strategies;* Water Environment Federation: Alexandria, Virginia.

Water Environment Research Foundation (2002) *Best Practices for the Treatment of Wet Weather Wastewater Flows;* Water Environment Federation: Alexandria, Virginia.

Water Environment Research Foundation (2005) *Identifying Technologies and Communicating the Benefits and Risks of Disinfecting Wet Weather Flows;* Water Environment Research Foundation: Alexandria, Virginia.

6

Facility Hydraulics and Pumping

Joseph C. Reichenberger P.E., BCEE, and David Terrill, P.E.

1.0 INTRODUCTION

Hydraulic considerations at wet weather treatment facilities are similar to other water resource recovery facilities (WRRFs) and each share the hydraulic principles presented in Chapter 6, "Plant Hydraulics and Pumping", of *Design of Municipal Wastewater Treatment Plants* (WEF et al., 2010). This chapter presents methods of controlling and splitting flows and discusses the advantages and disadvantages of typical methods. Influent flows to WRRFs originate from force mains, collection system interceptors, deep tunnels, or other conveyances and may or may not require pumping. Chapter 2 presented the regulatory background, characteristics of wet weather flows, and wet weather flow management; this chapter deals with the hydraulics of managing wet weather flows. Chapter 10 presents more detailed information on on-site wet weather storage and flow equalization facilities, which are mentioned in this chapter.

2.0 HYDRAULICS AND FLOW SPLITTING

2.1 Hydraulic Profile Effects

Chapter 6 of *Design of Municipal Wastewater Treatment Plants* (WEF et al., 2010) outlines WRRF hydraulic profile development, the need to provide adequate freeboard in unit processes, adequate freeboard below control weirs to avoid submergence, and sufficient hydraulic head (elevation difference) to permit gravity flow through each process under peak flow conditions. As mentioned in Chapter 6 of *Design of Municipal Wastewater Treatment Plants* (WEF et al., 2010), depending on expected surface disturbance and relative frequency of the condition, freeboard within tanks and channels can range from an extreme low of 150 mm to 1 m (6 in. to 3.3 ft) or more. Weirs are used for flow measurement and to maintain required water levels in process units. Freeboard below control weirs should be at least 50 mm (2 in.), preferably slightly more, to ensure free discharge (Benefield et al., 1984). This is good practice, but not always achievable under peak wet weather flow conditions.

During wet weather events, freeboard is often reduced and control and measuring weirs can become submerged. When weirs are submerged, flow measurement is unreliable. When clarifier weirs become submerged, effluent quality deteriorates, particularly if the weir scum baffles are overtopped and scum and floatables are discharged. Scum baffles in clarifiers should have sufficient height to be above the peak wet weather water surface elevation.

In addition to increasing plant influent flows, wet weather events frequently cause higher receiving water elevations. As indicated in Chapter 6 of *Design of Municipal Wastewater Treatment Plants* (WEF et al., 2010), the starting receiving water surface elevation during wet weather events is the projected maximum receiving water elevation for the event recurrence interval. This must be determined from flood maps or watershed hydrological studies. Often, gravity flow discharge is possible up to a certain facility flowrate and receiving water elevation. Gravity discharge should be maximized to reduce operating costs and energy consumption. Pumping or storage of effluent would be required when gravity flow is not possible. If pumping is required for effluent discharge, it is good practice to provide sufficient pumping capacity even with the largest pumping unit out of service.

If possible, the pumps should be on auxiliary or standby power to improve reliability.

Management of wet weather flows will likely require splitting the flows into parallel treatment trains and recombining flows into storage or equalization basins with repumping of stored flows for subsequent treatment. The following are options for flow splitting:

- Equal flow splitting to all parallel treatment process trains or
- Splitting the flow to not exceed the maximum flow that can be treated effectively by the main treatment process train and convey the remaining or excess flow to parallel treatment trains, storage, or equalization basins. The parallel train(s) would be designed to accommodate the anticipated peak flows.

The first option forces both the main and parallel process trains to respond to peak flows. This may not result in the best quality effluent. The second option maximizes the treatment efficiency of the main process train by limiting the flow and load fluctuations and transferring those fluctuations to the parallel process train, which can be designed to accommodate these flow and load variations. Treatment processes benefit from steady influent flowrates for efficient and economical treatment. Variable flowrates and influent constituent concentrations can cause significant fluctuations in dissolved oxygen control responses in aeration basins, sludge blanket depth in final clarifiers, short-circuiting in clarifiers, and disruptions to instrumentation and control systems, chemical feed systems, and other treatment processes. In an environment of ever restrictive National Pollutant Discharge Elimination System requirements, treatment processes do not need additional disturbance.

A variant of the second option is to gradually increase the flow to the main treatment process train with careful monitoring of process conditions and performance and adjusting the flow to the main treatment train up or down, as appropriate, as process conditions change. During this time, excess flows are diverted to separate wet weather treatment or storage facilities.

2.2 Hydraulic Fundamentals

Chapter 6 of *Design of Municipal Wastewater Treatment Plants* (WEF et al., 2010) provides a comprehensive discussion on treatment facility hydraulics and basic equations. These equations are applicable to wet weather treatment facilities. The basic pipe and channel conveyance, weir and orifice equations, and so forth are applicable. However, hydraulic performance of the specialized flow splitting equipment mentioned herein is best obtained from the equipment manufacturer. No attempt is made to provide this information in this chapter.

Before selecting a system or method for flow splitting, receiving water elevation and recurrence interval data should be obtained as this is typically the starting point for the hydraulic profile. The treatment facility hydraulic profile should be developed for various wet weather flowrates and recurrence intervals to determine if adequate freeboard exists in conduits and process units and flow measuring weirs are not submerged and to assess the effect of submerged clarifier and control weirs.

Based on the hydraulic profile study, an evaluation can be made of the available head for flow splitting systems. Some of the systems described in this chapter require significant hydraulic head to operate effectively. The total headloss at a weir includes the head above the weir plus the freeboard between the downstream water surface and the weir crest. This can total 0.35 to 0.5 m (1.25 to 1.6 ft) or more depending on the flowrate and weir crest length. The head may not be available and, hence, would eliminate that flow splitting option from further consideration.

2.3 Flow Splitting Options

Flow splitting can be passive or active. Passive systems require no operator intervention and have practical and operational advantages over active systems. Passive systems include vortex valves, orifices, constant level gates, or fixed overflow weirs. Weirs have the advantage of being relatively "clog free" from debris often associated with wet weather events. Unfortunately, except for vortex valves and constant level gates, most passive systems do not provide the same level of flow control that active systems can provide.

Active systems require manual operator intervention or some form of automatic control system to function. Active systems include, but are not limited to, manually or automatically actuated and modulating valves, motorized gates and folding weirs. For manual operation, an operating procedure that occurs sporadically can be challenging to operators because these procedures can be quite complex and variable from one wet weather event to the next, resulting from variations in the peak flow and its timing. Because it is not a matter of routine, this can lead to errors. Although mechanical and motorized flow splitting mechanisms such as modulating ports, weirs, gates, or valves can provide good flow control and flow splits, they require control systems that have all of the disadvantages inherent in any electrical and mechanical system (i.e., failure and high maintenance). Debris can compromise the operation of flow splitting measures. For example, butterfly valves do not function well in certain applications where large or stringy debris are present. Throttling valves are also subject to clogging. Downward opening gates are useful in areas where debris is significant. Upward opening gates do make good throttling devices because they can vibrate and become plugged with debris when they are in a nearly closed position. Valves should be selected to have a diameter large enough to pass the anticipated solids and have appropriate valve coefficient, C_v, across the entire range of anticipated flows to provide proper control. This is discussed later in this chapter.

Weirs, however, increase upstream hydraulic head, which may not be available. A constant level gate, when properly installed and sized, can maintain constant upstream water surface elevation. A side channel weir or a folding weir, placed adjacent to a properly sized open channel conveyance, can minimize diversion of bottom and bed load debris into a storage basin or parallel treatment stream. Orifice plates have practical limitations. They tend to over-restrict flows at low heads when not required, resulting in inefficient upstream hydraulic profiles and have increased potential for blockage. Vortex control valves provide the opportunity for adaptable flow regulation during wet weather events without the blockage problems presented by orifices.

2.3.1 Active Systems

Active systems include modulating weirs, tilting or folding weirs, modulating valves and sluice gates, and pumping systems.

2.3.1.1 Modulating Weirs

Weirs are classified according to the shape of the notch: rectangular, V-notch, trapezoidal, and proportional (Sutro) weirs. A Cipolletti weir is a special type of trapezoidal weir that has a sloping notch, typically 1 horizontal 4 vertical to account for "end contractions". Proportional or Sutro weirs have the discharge head varying linearly with flow. They are used in grit channels to maintain constant velocity with depth. Weirs are also classified as either "sharp-crested" or "broad-crested." Typically, broad-crested weirs are not used for flow splitting or flow control. For flow splitting systems, rectangular weirs are typically used. Proportional weirs would not typically be used in flow splitting applications.

Active systems have adjustable or modulating mechanisms, typically a hand-wheel (manual) operator or an electrical or hydraulic motorized operator, to raise or lower the weir crest to maintain a constant flow. The weir is typically on the side wall of the incoming channel or influent structure and can be upstream or downstream of the screening facilities. The weir discharges to a conveyance channel or pipe leading to a parallel wet weather treatment train or storage basin. Typically, a downward-opening, fabricated gate is used. Rubber seals on the gate minimize leakage. Because the flow over the weir is proportional to the depth of water over the crest of the weir, adjusting the crest of the weir adjusts the flowrate to the parallel treatment train or storage basin. The adjustment can be manual, that is, by operator intervention, or motorized based on a flow control loop designed to maintain a constant (or nearly constant) flow downstream to the main treatment process train. Typically, a flow meter is installed downstream of the flow splitting weir to measure the flow to the main process treatment train. In the motorized, automatic system, a programmable logic controller uses an adjustable flow set point in conjunction with the flow meter to maintain the preset desired maximum flowrate to the main process treatment train.

The system is effective at maintaining and controlling flow. Manual operating systems are not able to maintain as effective flow control as automatic, motor-operated systems without constant operator attention and intervention. Motor-driven systems are subject to wear in both the operating mechanism and the seals. Electric motor operators must be suitable for continuously modulating duty or they will fail. An advantage of using a downward opening weir gate is that there would be little or no effect on the water level in the incoming channel or sewer upstream of the flow splitting structure. The head calculation relationship for rectangular, downward opening gates can be calculated using the rectangular weir equation in eq 6.5 in Chapter 6 of *Design of Municipal Wastewater Treatment Plants* (WEF et al., 2010).

2.3.1.2 Tilting or Folding Weirs

Figure 6.1 shows a typical tilting or folding weir. The weir would typically be mounted on the side of the influent channel or structure and could be upstream or downstream of the screening facilities (downstream would be preferred). The weir discharges through a fabricated box into a conveyance channel or conduit leading to a parallel wet weather treatment train or storage basin. Adjusting the tilt controls the flow. Adjustment can be manual, automatic electric, or hydraulic motor operated. The tilting weirs can easily accommodate debris. As stated previously, manual operating systems are not able to maintain as effective flow control as automatic, motor-operated systems and they are subject to wear in both the operating mechanism and the seals. Electric motor operators must be suitable for continuously modulating duty. An advantage of using a tilting weir gate is that there would be little or no effect on the water level in the incoming channel upstream of the flow splitting structure. Although these tilting weirs are "rectangular" weirs, they have a sloping approach channel that would make the rectangular weir equation (see, eq 6.5 in Chapter 6 of *Design of Municipal Wastewater Treatment Plants* [WEF et al., 2010]) inapplicable. The design engineer should consult the manufacturer for flow-head data.

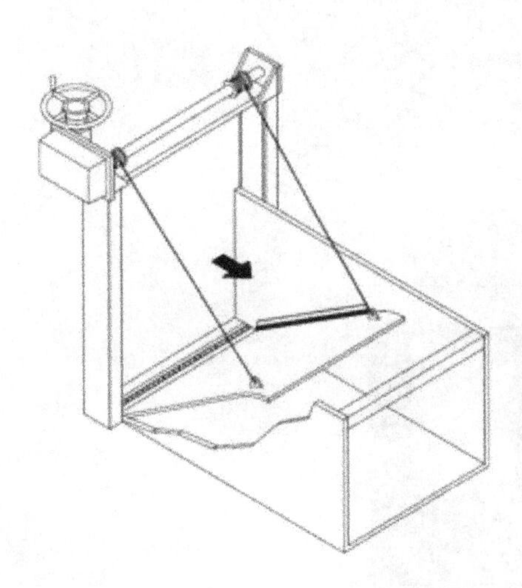

(courtesy of Waterman Industries)

(courtesy of Martin Childs Ltd.)

FIGURE 6.1 Tilting or folding weirs.

2.3.1.3 Modulating Valves and Gates

Almost any type of valve could be used to control flow splits. Common types include plug, cone- and ball-type valves, butterfly valves, and "pinch" valves. Gate valves should be avoided because they have poor throttling characteristics. Plug and cone and ball valves perform adequately in wastewater applications, but are expensive in larger sizes. Butterfly valves and butterfly gates (available in rectangular configurations) are subject to clogging from stringy materials common in wastewater. Butterfly valves would perform satisfactorily with screened and settled wastewater; as such, they could be used to control flow from storage basins and tanks. Pinch or weir valves have a flexible rubber liner that can be squeezed by a compression actuator (see Figure 6.2). These valves have full port openings and are not subject to clogging unless operated with only a small opening. Should partial clogging occur, a downstream flow meter in the control loop would sense the decreased flowrate and open the valve, allowing the debris to pass through. Sometimes on low-pressure (low-head) operation, the rubber sleeve may not be sufficiently rigid and may tend to flap while operating, affecting the flow control. The designer should consult the manufacturer for specific applications. The headloss through valves can be determined using eqs 6.14 through 6.16 in Chapter 6 of *Design of Municipal Wastewater Treatmet Plant* (WEF et al., 2010).

The value for the valve coefficient, K_v or C_v, in the aforementioned referenced equations varies with the "percent open" of the valve. The valve size should be selected to ensure good control over the entire range of anticipated flow.

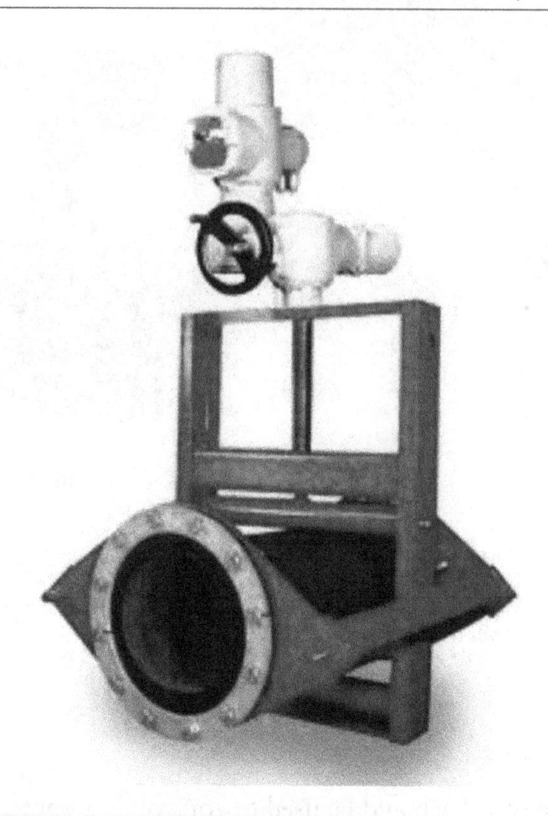

FIGURE 6.2 Motor-actuated pinch valve (courtesy of Red Valve Company).

A sluice gate or fabricated gate could be used in a flow splitting role; they are typically used where larger flows are anticipated. The gate typically opens at the bottom and allows flow under the gate to either enter the main process treatment train or a parallel treatment train or storage basin. Either gate can be manually or continuously modulating using an electric or hydraulic motor in conjunction with a flow meter similar to the control for the weir gates described previously. Because they open at the bottom, hydraulically they operate as orifices and the hydraulic losses are calculated using the orifice equation (WEF et al., 2010). The orifice headloss could affect the upstream water surface elevation. Upward opening gates should be used with caution because they are not typically designed for throttling. High-velocity flow under the gate will cause vibration and possibly cavitation, which might result in premature gate failure.

2.3.1.4 Pumping

With a pumped system, excess flow will be pumped from the facility influent sewer or wet well by a variable-speed or constant-speed pump to a parallel treatment train or storage basin. Using a pump system allows the location of the storage basin or parallel treatment train to be located at a higher elevation, which could permit gravity return flow from either the storage basin or the parallel treatment train. Although gravity flow splitting to a storage basin is desirable using one of the other active systems described previously, that is not always possible. Gravity flow has some advantages in energy savings and improved reliability because diversion to storage basins can occur uninterrupted even during power outages. If gravity inflow to a storage or equalization basin is possible, pumping of the stored contents to the treatment facility is typically required. When pumping into a storage basin, the pumps must be sized to cover the full range of anticipated flows. When pumping out of storage basins, the pumps can be designed for a lower pumping rate with reduced costs. Pump systems that are properly designed should not affect the water surface elevation in the incoming conduits or channels and, therefore, not affect the upstream water surface in the influent sewer and conduits.

A typical system could have a manually or automatically actuated sluice gate in the incoming sewer, channel, or main pumping station that would open fully to divert flow into a separate wet weather pumping station. Pumps in the wet weather pumping station would lift the flow to the parallel treatment train or storage basin. Pumps, which lift wastewater to a parallel process train, should be variable speed so that flow changes are gradual. Pumps that lift wastewater to a storage basin could be constant speed, but, if they are, the effects of flow transients caused by stopping and starting the constant-speed pumps on the main treatment process train should be evaluated.

2.3.2 Passive Systems

Passive systems include vortex valves, orifices, constant level gates, or fixed overflow weirs. Fixed overflow weirs can be perpendicular to the flow or installed in the side of conveyance channels (side channel weir or spillway). Side channel weirs are complicated hydraulically and controlling the flowrate being split off is difficult; therefore, this will not be discussed in this chapter. The approach velocity and end contraction effect in fixed weirs installed perpendicular to the flow must be carefully evaluated in the design hydraulics.

2.3.2.1 Vortex Valves

A vortex valve consists of an intake opening, a volute, and an outlet (see Figure 6.3). Under low water level conditions at the inlet, wastewater flows tangentially into the inlet and volute and out through the outlet. As the head or water level upstream of the vortex valve increases, a circular flow pattern

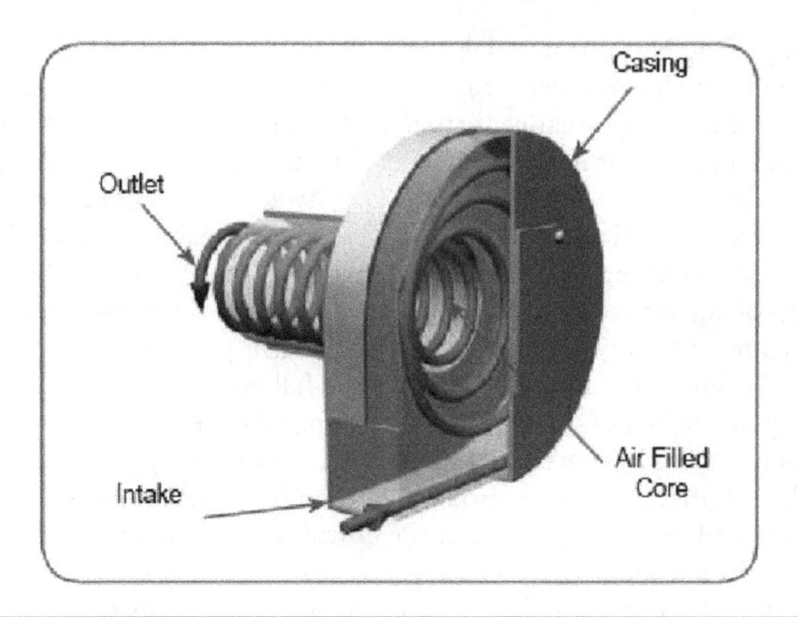

FIGURE 6.3 Vortex valve (courtesy of Hydro-International).

develops in the valve that induces a vortex in the control opening. This creates an air core that restricts the flow out through the outlet. The greater the head upstream, the larger the air core and the greater the restriction. Vortex valves can provide a significant reduction in peak flowrate without any mechanical intervention. Because the vortex valve has a larger outlet opening than an equivalent circular orifice producing the same headloss, vortex valves are less subject to clogging.

Vortex valves are best installed at the outlet of storage basins to control the gravity flow out of the basin into the treatment process. If a vortex valve is installed in the influent channel or conduit to the treatment facility, a side overflow weir would be needed to discharge excess flow to a parallel treatment train or storage basin. Vortex valves develop significant headloss and, as such, would raise the upstream water surface if installed in a conduit or channel. This may not be acceptable.

2.3.2.2 Orifices

Orifices are fixed-size openings designed to regulate flow, typically to some predetermined maximum flowrate based on the available head. Orifices can provide effective flow control, but have practical limitations. To regulate flow effectively, significant hydraulic head is required. Orifices are subject to clogging and, therefore, may not be the best choice to accommodate wastewater or wet weather flows. Orifices could be used to limit the rate of flow of screened and settled wastewater from tanks and storage basins, but clogging is still possible. Hydraulic losses for orifices are calculated using the orifice equation (WEF et al., 2010).

2.3.2.3 Constant Level Gate

Figure 6.4 shows a schematic of a constant level gate. These gates are designed to maintain a constant upstream or downstream water level. In a flow splitting operation, the upstream water level can be maintained at a constant level matching the water surface elevation at the maximum desired flow through the main process train. The gate underflow would go to a parallel process train or storage basin. One manufacturer offers an adjustable level gate wherein the constant upstream level can be manually adjusted to suit conditions. Headloss information is best obtained directly from the manufacturer.

2.3.2.4 Fixed Overflow Weir

Fixed weirs (rectangular, Cipolletti, triangular) offer the advantage of low maintenance and simplicity. Flowrate over the weir is an exponential function of the depth of water above the weir. Maintaining close control on the maximum amount of wastewater through the main process treatment train requires a lengthy weir, particularly if flows are large. This space is often not available. If weirs are too long, flow splitting accuracy decreases. If a shorter weir is used, the weir crest may have to be set at a lower elevation than is optimum and, as a result, flows may be sent to storage basins or parallel treatment trains before the main process train reaches maximum capacity. Headloss eqs 6.4 and 6.5 in Chapter 6 of *Design of Municipal Wastewater*

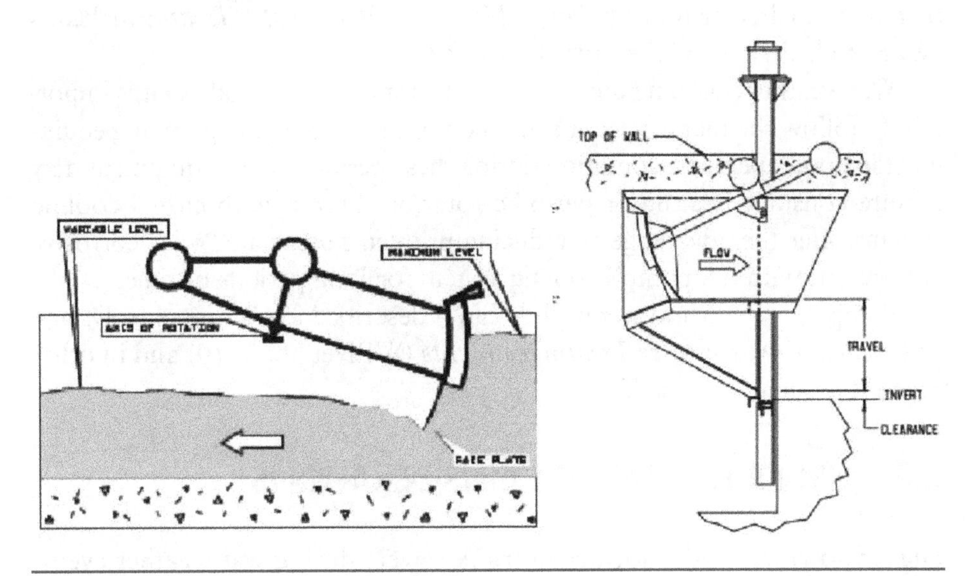

FIGURE 6.4 Constant level gate (courtesy of Waterman Industries).

Treatment Plants (WEF et al., 2010), are applicable to triangular (V-notch) and rectangular weirs, respectively. Equation 6.5 in Chapter 6 of *Design of Municipal Wastewater Treatment Plants* (WEF et al., 2010) is also applicable to Cipolletti weirs, where the weir crest length is measured at the bottom of the Cipolletti weir. End contraction corrections are not required when using a Cipolletti weir.

3.0 PUMPING DESIGN

The designer has two options to consider: fixed- (constant-) speed or variable-speed pumps. Variable-speed pumps offer the advantage of gradually increasing or decreasing flows over the pump's range of operation, which has advantages to downstream processes because the flowrate changes over time are more gradual. Constant-speed pumping into a storage or equalization basin would be appropriate, however, because the on–off cycling would not affect the contents of the storage or equalization basin. Variable-speed pumps typically have a turndown capability to approximately 40 to 50% of their design capacity. The actual turndown capacity depends on the system head curve, the shutoff head, and the shape of the pump performance curve. As such, a combination of smaller and larger pumps may be necessary to accommodate the desired flow range. Because variable-speed pumps can operate at lower flows, wet well storage volumes between pump starts can be reduced from those required for constant-speed pumps. The wet well volume between pump start-and-stop levels depends on the minimum cycle time (i.e., time between pump starts), motor size and type, and the pumping rate at startup. The calculation procedure can be found in Chapter 6 of *Design of Municipal Wastewater Treatment Plants* (WEF et al., 2010) and literature by Jones (2006).

Wet weather pumping units may sit idle for certain periods so it is important to follow the manufacturer's instruction for keeping the pumping equipment in good operating condition during these periods. Some equipment may require constant heating or periodic rotation. Pumps with closed cooling systems offer the advantage of reducing problems associated with corrosive effluent left within a pump's cooling system for long periods of time.

Pumping system and wet well design is described in Chapter 6 of *Design of Municipal Wastewater Treatment Plants* (WEF et al., 2010) and in other references such as Jones (2006).

4.0 OTHER HYDRAULIC CONSIDERATIONS

Higher flowrates and velocities in trunk sewers during wet weather events scour debris and solids deposited during dry periods and convey them to the WRRF where they can collect in wet wells and conveyance channels. Provisions should be made in the design to facilitate removal of debris from wet wells.

Conveyance channels and conduits must maintain adequate velocities to transport the solids to the grit removal process. Sharp, 90-deg. channel corners should be avoided because these locations will accumulate grit and solids. Smooth radius corners and filets should be considered. Grit chambers, grit pumps, and grit handling equipment must be designed for the increased load during wet weather events.

Additional bar screens may be needed during wet weather events to accommodate the increased amount of debris and avoid excessive velocity through the screens. Excessive velocities not only cause additional headloss, but also reduce screening system performance.

5.0 ODOR CONTROL

Chapter 7 of *Design of Municipal Wastewater Treatment Plants* (WEF et al., 2010) and *Control of Odor and Emissions from Wastewater Treatment Plants* (WEF, 2005) provide guidance for odor control systems and odor control strategies. Treatment processes for wet weather flows, such as pumping stations, screens, and clarifiers, and wet weather storage basins can be a source of odor, particularly if not cleaned immediately after use. If wet weather flows are to be stored for extended periods of time, supplemental aeration or covering of the tanks and basins may be necessary.

6.0 REFERENCES

Benefield, L. D.; Judkins, J. F., Jr.; Parr, A. D. (1984) *Treatment Plant Hydraulics for Environmental Engineers;* Prentice-Hall: Englewood Cliffs, New Jersey.

Jones, G. M., Ed. (2006) *Pumping Station Design,* 3rd ed.; Butterworth-Heinemann: Burlington, Massachusetts.

Water Environment Federation (2005) *Control of Odor and Emissions from Wastewater Treatment Plants;* WEF Manual of Practice No. 25; Water Environment Federation: Alexandria, Virginia.

Water Environment Federation; American Society of Civil Engineers; Environmental & Water Resources Institute (2010) *Design of Municipal Wastewater Treatment Plants,* 5th ed.; WEF Manual of Practice No. 8; ASCE Manuals and Reports on Engineering Practice No. 76; Water Environment Federation: Alexandria, Virginia.

7

Support Systems

Vamsi Seeta and Jann Yamauchi

1.0 INTRODUCTION

As important as the core technologies/processes to handle additional flows from wet weather events at water resource recovery facilities (WRRFs), support systems that keep the facilities in operation are equally critical for their reliability. Some of the key support systems included for discussion in this chapter are electric power, standby power, chemical feed and storage facilities, and communications and controls.

2.0 POWER SYSTEM

In instances of extreme wet weather events, catastrophic failures may occur, leading to a complete loss of utility power. Depending on the nature and extent of failure, this could be a short-term, temporary situation, or could lead to extended periods of blackout. There are several planning and operational strategies developed that tend to mitigate the negative effects of such catastrophic failure on WRRFs.

2.1 Power Supply Reliability and Redundancy

Reliability and adequate redundancy are key features of the power system at WRRFs. To meet these criteria, many facilities have installed two separate and independent electrical utility power sources. Where independent feed sources are not feasible, facilities have opted to install an on-site power generator for additional reliability to the power system. For facilities that rely on a single electrical feed to supply the power and lack additional on-site reliability, wet weather events pose greater risk of complete loss of power.

Adequate capacity of the feeder lines is another key component of operational strategy for WRRFs. In instances of redundant feeder lines, where each feeder is not capable of supporting the full connected load at the facility, operators need to be aware of a possible "load shedding" protocol if one of the feeders is affected by wet weather events.

It is also important to plan for redundancy at multiple levels in the power system, starting with utility transmission lines, utility substations, facility-side distribution lines, and transformers. Some of the key methods used to avoid a common mode of failure is to build sufficient redundancy by installing independent sources of power for transformers and double-ended motor control centers with automatic power transfer schemes and reliable standby power generators.

2.2 Demand-Side Management

2.2.1 Power Demand Spikes

Wet weather events typically exert hydraulic overload on the WRRF processes. Depending on how each WRRF is set up to handle the additional volume of wastewater, there could be a potential for electrical load surges. These "spikes" in power demand are a result of bringing additional equipment online in a short timeframe. Most utility companies across the country charge municipalities based not only on the total power consumed, but also on the "peak" power demand, even if it lasts only for a short duration. These peak demand charges could be significant and each facility should determine their operational strategy to "manage" the demand and avoid or minimize such charges.

2.2.2 *Load Management—Peak Load Shifting/Load Shedding*

Electrical load management is an effective method to mitigate negative effects of sudden spikes in power demand during wet weather events. Wet weather storage facilities could be used to store the additional volume until after the utility company defined "peak periods". The stored wastewater could then be treated during offpeak hours. Another strategy used is to operate a "peak shaving" facility that operates on an alternative fuel source, such as anaerobic digester gas-powered internal combustion engines, to offset the power imports during peak periods. These strategies will not only avoid the risk of sudden power surges, but could also result in significant savings from reduced/avoided demand charges from the utility company.

Further, biosolids facilities are typically designed and constructed with a considerable amount of excess capacity and often run only a few hours a day. During wet weather events, processes such as tertiary treatment and biosolids thickening and dewatering could be stopped or operated with deferred schedules to minimize the effects on the facility power system. Thus, effective load management during wet weather events involves implementing a combination of shifting the operations of processes to offpeak hours and shedding a portion of the total load by not operating the processes that are not critical to facility performance.

2.3 Inrush Current and Mitigation Measures

Typically, regular alternating current induction motors require high starting (inrush) current, as much as 6 to 8 times their full load amperage draw, to maintain high efficiency and power factor, among other factors. As discussed earlier, because of the spikes in power demand when equipment responds to wet weather, it is possible to have a huge surge in power draw by the motors that are connected across the line. This has a potential to adversely affect the voltage across the facility, thereby affectiing other electrical equipment connected to the same electrical system. Several mitigation measures could be deployed to reduce the overall inrush current when starting large motors such as wet weather pumps.

Solid state drives help regulate the power delivered to the motor by controlling the voltage. This facilitates in a "soft" start by reducing the inrush power draw. However, because motors require a minimum voltage to generate adequate torque for operation, the solid state drives can only be used for limited voltage control. On the other hand, variable frequency drives (VFDs) achieve the inrush current mitigation by regulating the frequency input to the motor. The power draw of the motor with a VFD will never exceed the full load amperage rating of the motor, thereby avoiding the negative effects of an inrush current situation. Variable frequency drives also provide the added advantage of maintaining higher power factor, which might avoid penalties from the utility company based on "poor" power factor.

Another advantage with VFDs is that they mitigate harmonic distortions in the electrical system, which can interfere with sensitive electronics and communication systems as well as electrical equipment such as transformers. Some utilities penalize facilities for introducing harmonics on their grid, while incentivizing the efforts to mitigate them. Several methods are used in the industry to mitigate harmonics; the following methods are more commonly used:

- Line reactors,
- Passive and active filters, and
- Multipulse VFDs.

Variable frequency drives used on a 3-phase alternating current in the United States have a minimum of six rectifiers, indicating the number of pulses. The more the number of pulses, the better the harmonics mitigation effect. Currently, manufacturers offer 12-, 18-, 24-, and 30-pulse drives.

3.0 STANDBY/EMERGENCY POWER SYSTEM

3.1 Critical Loads

Critical loads at a WRRF are those loads that have an absolute necessity to be in operation at all times, especially during wet weather events when the facility loses utility power. Typically, critical loads are identified from a facilitywide motor load list that represents the power distribution system. This load list describes the criticality of each connected load and classifies each load based on its operation schedule as "continuous duty", "intermittent duty", or "standby duty". Typically, the capacity of on-site emergency/standby generation capacity is determined based on the critical loads that it needs to support considering a total power loss scenario during wet weather events. Examples of critical loads at a typical WRRF include influent wastewater pumps and controls, any intermediate lift pumps, primary and secondary clarifier drives, aeration blowers and effluent disinfection (chlorination), and dechlorination facilities.

3.2 Diesel Emergency/Standby Engine Generators

It is a common practice at WRRFs to use diesel-fueled internal combustion engine generators as the emergency/standby power system. Although there are other forms of backup power systems available, diesel engine generators provide unmatched reliability when the utility power is lost as a result of extreme weather conditions. For instance, natural gas fueled engine generators are also popular in the industry to provide emergency/backup power. However, natural gas pipelines could be vulnerable to damage during catastrophic events and may render the natural gas based emergency/standby power systems inoperable. On the other hand, diesel fuel is required to be stored on-site in underground storage tanks, ready to be used when needed. Facilities need to ensure that the diesel refueling schedule is properly planned out and coordinated with suppliers to ensure that an adequate quantity of diesel fuel is stored on-site at any given time.

Load shedding is an optional control feature of emergency/standby engine generator systems that trips predesignated loads while switching power from a utility company to an on-site power system. By reducing the loads that need to be started initially, the sizing of the emergency/standby power system could be reduced. Further, "block" loading the engine generator with all the loads at the start time could lead to variations in voltage and frequency output, thereby reducing the effectiveness of the on-site backup system. A load sequencing plan could be programmed to ensure that once the engine generator is operational, additional loads can be connected sequentially.

3.3 Diesel-Fueled Engines and Pumps

As an alternative to "utility powered" critical equipment such as stormwater pumps, using diesel-powered engines that drive the pumps and other wet weather equipment could provide distinct advantages during wet weather events. By switching the prime mover from an electric motor to a diesel-powered engine, the same reliability could be achieved as in instances of diesel-fueled emergency/standby engine generators. In addition, by selectively choosing the equipment that are large electrical loads and converting them to diesel fuel based will reduce the peak load on the emergency/standby engines and result in a more stable electrical system during startups.

3.4 Co-Generation Engines as Backup Power System

Digester gas and/or natural gas powered internal combustion engine generators used for combined heat and power applications at a WRRF can also be used as an effective backup power system. These power generators could serve as supplemental/additional backup sources or could be used in lieu of on-site diesel generators. If these gas-powered engine generators are used to provide backup power during wet weather events, care should be taken to store adequate amounts of gas to start the engines in case of utility power failure. This feature, commonly known as "black start", requires a properly sized battery backup system to start the engine controls.

4.0 CHEMICAL FEED AND STORAGE FACILITIES

Chemical facilities could play a vital role in properly handling wet weather flows at a WRRF. Chemically enhanced primary treatment could be used as a full-time treatment method at a WRRF, or it could be deployed exclusively during periods of wet weather to handle peak flowrates.

Water resource recovery facilities need to plan the chemical bulk storage, preparation, and feeding facilities to handle additional influent flows during wet weather events. Proper planning for these facilities takes into account the minimum and maximum chemical dosages for normal operation with appropriate allowance for wet weather flows. Planning should also consider potential difficulties with logistics during severe wet weather events, which could affect the "normal" procurement process for chemicals. However, WRRFs that use chemicals only during wet weather events need to install dedicated chemical dosing systems, which should be properly maintained so that they will perform as intended when such events occur. Considerations should include the type of chemical used, specific deterioration rate, commercially available delivery loads, lead times for procurement, and the supplier's responsiveness for chemical delivery requests. Proper care must be taken to periodically inspect and check for the integrity of the chemicals and a plan is required for fresh chemical deliveries well in advance to ensure chemical availability on-site at all times. Chemical dosing pumps need to be maintained per manufacturer recommendations and dosing pipes need to be inspected for chocked conditions. Depending on the weather conditions and chemicals being handled, heat tracing of chemical tanks and dosing pipelines should be considered.

5.0 COMMUNICATIONS AND CONTROLS

Communications and controls are at the heart of WRRF operations. The use of centralized computer controls such as supervisory control and data acquisition systems in WRRFs is expanding rapidly, even to smaller facilities. The centralized controls integrate existing stand-alone controls or distributed control systems, improving operational efficiency and facilitating monitoring of response from field-mounted instruments during wet weather events.

Centralized control systems allow for integrated data collection and analysis and provide opportunities to improve overall facility performance. Supervisory control and data acquisition systems at WRRFs provide key information to direct when to operate remote equipment and make complex decisions based on input from the system. Supervisory control and data acquisition systems provide continuous and precise control of process variables and can start, slow down, or stop equipment in response to feedback acquired from field-mounted instruments. These systems provide real-time demand control and allow facility managers to coordinate and schedule a variety of activities in response to wet weather conditions, such as load shedding and shifting and bringing equipment such as storm water pumps online and offline.

Equipment such as gates and valves that are manually placed into operation only during wet weather events may not function properly when most needed, as a result of lack of frequent maintenance. Such critical flow diversion/control equipment should be equipped with "fail-safe" hydraulic actuators to ensure quick and proper response. In instances where appurtenances are controlled using electric operators at critical facilities, manual overrides should be installed with hand operators. Clear operation protocols need to be developed to identify systems equipped with fail-safe operators and those that require manual intervention and to provide guidance to operators with step-by-step instructions.

6.0 SUGGESTED READINGS

Water Environment Federation; American Society of Civil Engineers; Environmental & Water Resources Institute (2010) *Design of Municipal Wastewater Treatment Plants*, 5th ed.; WEF Manual of Practice No. 8; ASCE Manuals and Reports on Engineering Practice No. 76; Water Environment Federation: Alexandria, Virginia.

Water Environment Federation (2013) *Guide to Municipal Wet Weather Strategies;* Water Environment Federation: Alexandria, Virginia.

Water Environment Research Foundation (2002) *Best Practices for the Treatment of Wet Weather Wastewater Flows;* Water Environment Research Foundation: Alexandria, Virginia.

8

Occupational Health and Safety

Stacy J. Passaro, P.E., BCEE, and Alan Scrivner

1.0 SAFETY DURING WET WEATHER EVENTS

There are many safety hazards that are present when a person is working at a water resource recovery facility (WRRF). This is the result of the huge range of activities that facility operations and maintenance personnel must carry out. In the United States, the relative risk and danger of working in the wastewater industry is tracked by the Bureau of Labor and Statistics (BLS). Based on 2009 injury/illness/fatality (IIF) data that were collected from 41 states, BLS determined that the IIF incident rate in the wastewater industry was 4.1 incidents per 100 workers, while the overall national average for all industries had an average of 3.6 IFF incidents per 100 workers. Historically, the wastewater industry incident rate has been over 5 incidents per 100 workers.

The following categories show the types of safety incidents commonly experienced at wastewater facilities:

- Trips and falls,
- Strains and sprains,
- Electric shock/arc flash,
- Stored energy,
- Chemicals,
- Pathogens, and
- Confined spaces.

One of the biggest challenges to achieving improved safety ratings in WRRFs is the breadth and scope of activities that operations and maintenance personnel must participate in while conducting their day-to-day job duties. Being aware of hazards in the workplace, gaining a well-rounded knowledge of the sources of those hazards, and consistently applying risk mitigation techniques are all part of a successful strategy to manage these risks. These risks can also be exasperated by the fact that facility personnel must sometimes react to situations that occur during shifts that may have reduced staffing levels. Table 8.1 contains Web site links to additional sources of safety, health, and security information and risk mitigation.

Water Environment Federation (WEF) Manual of Practice No. 1 (WEF, 2013), titled *Safety, Health, and Security in Wastewater Systems*, includes detailed information regarding how to establish or expand a site-specific

TABLE 8.1 Safety, health, and security reference table.

Topic	Agency/organization	Web site
Federal safety regulations	OSHA	http://www.osha.gov/
State OSHA plans	OSHA Web page link to state plans	http://www.osha.gov/dcsp/osp/index.html
Injury and illness prevention	National Institute for Occupational Safety and Health	http://www.cdc.gov/niosh/
Traffic safety	National Highway Traffic Safety Administration	http://www.nhtsa.gov/
Chemical safety	U.S. Chemical Safety Board	http://www.csb.gov/
Public facility security	Department of Homeland Security	http://www.dhs.gov/index.shtm
Safety industry	American Society of Safety Engineers	http://www.asse.org/
Fire protection	National Fire Protection Association	http://www.nfpa.org/
First aid and emergency response	American Red Cross	http://www.redcross.org/
Work, home, and community safety	National Safety Council	http://www.nsc.org/Pages/Home.aspx
Injury and illness statistics	Department of Labor, Bureau of Labor and Statistics	http://www.bls.gov/iif/
Hazardous materials transportation safety	National Transportation Safety Board	http://www.ntsb.gov/

safety and health program. In general, the following components must be addressed:

1. Commitment to the program from management,
2. Employee participation in development of the program,
3. Hazard identification,
4. Hazard prevention and control,
5. Training and verification of training effectiveness, and
6. Ongoing evaluation of the safety program.

Section 2.0 of this chapter is geared to helping identify and mitigate safety risks at WRRFs. Water Environment Federation's *Safety, Health, and Security in Wastewater Systems* (WEF, 2013) presents additional information on general safety risks and establishing health and safety programs. The rest of this chapter will deal with identifying and addressing safety risks related to WRRFs and related assets during wet weather events.

2.0 WASTEWATER WET WEATHER INFRASTRUCTURE AND SPECIFIC HAZARDS

Hazard identification, prevention, and control are important steps in developing an effective strategy to reduce safety and health risks at a WRRF during wet weather events. This section reviews many of the common wet weather risks to help with the identification step of the process. A WRRF should review all unit processes and assets and develop a list of risks that are present during wet weather events. After the risks are identified, it is easier to research the potential mitigation strategies and to educate all facility personnel about the hazards to identify a means to avoid the hazard or to reduce the risk to an acceptable level.

2.1 Manholes

Manholes can become surcharged during wet weather events. Unrestrained covers, which are typically more than 45 kg (100 lb), have been blown off of the manhole by slugs of water and thrown 15 to 18 m (50 to 60 ft) through the air. Restrained lids have created high-pressure conditions in which a sudden failure of the side of the concrete structure has occurred and chunks of concrete, rebar, and manhole covers and frames have "exploded" apart.

2.2 Wet Wells and Dry Wells

Wet wells can fill rapidly and overflow. Dry wells can become flooded in extreme wet weather events. Because these areas are not intended to have water present, electrical hazards may be present in a flooded condition.

2.3 Vaults

Vaults can hold critical equipment that is located below grade that may need to be accessed. Many vaults can become flooded under high groundwater or surface flooding conditions.

2.4 Force Mains and Appurtenances

Force mains can be exposed and undermined if they are located in areas where erosion resulting from flooding may occur.

2.5 Screens

Stationary bar screens and bar racks will require much more frequent inspection and manual cleaning to ensure that debris is not blinding the device. Mechanical screens that typically operate intermittently may be required to run constantly to remove incoming debris. The beginning of an intense wet weather event may carry a large quantity of debris because of "first flush" effects. Screens will require more frequent inspections to ensure that the screens are not becoming blinded with debris, that channel water elevations are not in danger of topping the concrete walls, and that dumpsters receiving screenings are not overfilling.

2.6 Grit Removal Systems

The beginning of an intense wet weather event may carry high grit loads because of first flush effects. This equipment needs to be monitored more frequently than is typical to ensure that the unit is not overloaded and that dumpsters are not full or overflowing.

2.7 Flow Channels

As flows increase, velocities in channels will increase as well. Facility operations and maintenance personnel need to be aware that velocity currents through these channels will be much higher than normal.

2.8 Equalization Tanks/Basins

Equalization tanks often sit empty, sometimes for significant periods of time. If these tanks must be inspected in anticipation of a wet weather event, all applicable safety precautions such as confined space entry procedures must be followed.

When wet weather occurs, excess peak flows may be diverted from the process to these holding tanks to protect the treatment train equipment and process. Pipes, channels, and basins that are typically empty may quickly be put into use with little or no advance warning. Personnel and equipment should not be in these tanks when there is a threat of rain or thunderstorms. Weather forecasts should be monitored frequently, particularly in areas subject to isolated local storms.

2.9 Process Tanks

Process tanks are a common feature at a WWRF. Because they are often used in various ways, they may be in service or out of service and the water level in each could be higher or lower depending on the process being conducted in that tank.

2.9.1 Flooded Conditions

Flooded conditions can occur quickly during a wet weather event. This can be made worse when loss of electrical power also occurs.

2.9.2 Empty Tanks

During wet weather events, groundwater table elevations can rise quickly. If tanks are empty and are not properly anchored and vented, groundwater can exert pressure on the tank and cause the tank to move or "float". This can cause pipes to shear and concrete to crack. Tanks that are moved by groundwater pressure can be seriously damaged and typically must be completely removed and replaced. If items like fuel lines or chemical lines are damaged during the tank lifting, extremely dangerous conditions can occur. If these tanks need to be entered for any reason including an emergency inspection, all applicable safety precautions such as confined space entry must be followed.

2.10 Emergency Process Modes

To protect the biological treatment process, many facilities will temporarily switch to emergency process modes. The required steps to change the facility over to the emergency mode should be written out. Issues should be planned for such as opening or closing gates or valves and moving stop logs or metal stop gates with and without power.

2.11 Chemical Feed for Wet Weather Treatment

Some WWRFs will use chemicals such as polymer, alum, or ferric chloride during wet weather events to improve short-term performance of secondary clarifiers, effluent filters, or chemically enhanced primary clarifiers. Facility personnel need to be aware of the safety concern related to these chemicals that are intermittently used. Training on proper handling of these chemicals needs to be conducted and operators need to be aware and cautious in these areas as use of the chemicals is not the everyday, typical practice.

2.12 Satellite/Remote Wet Weather Treatment Facilities and Filters

These systems may be out of service for weeks or months between wet weather events, which is when they are needed. Inspections and preoperational checks should be made before an anticipated wet weather event. Many satellite treatment systems will automatically start up when certain conditions such as sewer surcharging occur.

2.13 Power Systems

During wet weather events, care must be taken to avoid personal injury resulting from electrical shocks, arc flash issues when accessing electrical equipment, and other electrical dangers. Emergency power generation systems are often used during wet weather events because of utility power losses caused by wind damage or flooding issues. Care must be taken in transferring power safely from one source to another. Chapter 7 includes detailed information about electrical system operation and potential issues during wet weather events.

2.14 Access Road Flooding

Because sewer systems are designed to flow by gravity where possible, a WRRF is typically located at a low-lying elevation. Because effluent is discharged to a local waterbody, facilities are typically located immediately adjacent to a lake, river, creek, or other waterbody. To rectify conditions that have led to previous flood events, many facilities have built berms around the facility in an attempt to keep floodwaters outside of the facility's fence line. It is imperative that advance planning takes place to ensure that personnel, who may become trapped inside the facility by floodwaters, have adequate supplies such as food and clean water. Motorized vehicles should never be driven through standing water to attempt to reach or leave the WRRF. Some facilities keep a boat on-site to be used in emergency flooding situations to get people in or out as needed. Extra care must be taken to navigate a boat safely though floodwaters.

2.15 Building Flooding

Flooding can cause electrical hazards; damage critical pumps, instrumentation, and electrical/controls equipment; and lead to chemical spills.

2.16 Site Construction Effects

Construction tends to disrupt the site's regular drainage patterns and normal operation of the site stormwater system. Unusual flooding dangers are more likely to occur under these conditions.

2.17 Climatic Effects

Severe weather such as lightning, hail, snow, wind, thunderstorms/supercells, tornados, and hurricanes can create sudden dangerous situations. Trying to keep a WRRF in operation or at least to minimize damage through a severe weather event can create many hazardous conditions. Regular checks should be made on the grounding loops. Flooding issues and emergency power arrangements need to be planned for ahead of the event.

2.18 Summary

After a WWRF identifies all of the site-specific hazards that exist at the facility, a prioritized list of actions that can be taken to mitigate these risks must be developed. Emergency procedures must be established in writing and regular training must be given to ensure that all personnel are well-versed in implementing these procedures should the need arise.

The Office of Safety and Health Administration (OSHA) has developed several resource documents that provide additional safety information that apply to wet weather operations and recovery. The name of the document and a link to the OSHA resource are given in the following list:

- Electrical hazards: https://www.osha.gov/dts/weather/flood/response.html#electrical
- Tree and debris removal: https://www.osha.gov/dts/weather/flood/response.html#debris
- Carbon monoxide: https://www.osha.gov/dts/weather/flood/response.html#carbon_monoxide
- Lifting injuries: https://www.osha.gov/dts/weather/flood/response.html#lifting
- Mold: https://www.osha.gov/dts/weather/flood/response.html#mold
- Chemical and biological hazards: https://www.osha.gov/dts/weather/flood/response.html#chemical
- Fire: https://www.osha.gov/dts/weather/flood/response.html#fire
- Drowning: https://www.osha.gov/dts/weather/flood/response.html#drowning
- Hypothermia (resulting from cold weather and water exposure): https://www.osha.gov/dts/weather/flood/response.html#hypothermia
- Exhaustion (from working extended shifts): https://www.osha.gov/dts/weather/flood/response.html#exhaustion
- Heat: https://www.osha.gov/dts/weather/flood/response.html#heat

3.0 WET WEATHER RISK MANAGEMENT

A systematic approach to risk management at a WWRF will include special circumstances that occur during wet weather events. The previous section outlined some of the hazards associated with a wet weather event. This section discusses how hazards specific to a WRRF are analyzed and the risks of an accident are minimized.

3.1 Risk Analysis

The traditional approach to preventing accidents that cause physical injuries involves recognizing, evaluating, and controlling hazards. A *hazard* is defined as the capacity to cause harm and is a quality of a material or a condition. A toxic chemical such as hydrogen sulfide or chlorine gas has the capacity to cause harm. *Risk* is the chance or probability that a person will experience harm. Risk is not the same as a hazard. Risk involves both probability and severity elements. The hazard of chlorine gas is the same regardless of where it exists. The probability of being injured by the chlorine gas is significantly reduced if the chlorine gas is stored in a separate, well-ventilated room rather than in the middle of a process area.

During a wet weather event, risks and hazards are much greater than during normal dry weather flows. In a wet weather event, more hazards exist for some of the following reasons:

- More equipment is in use creating greater opportunity to contact moving or energized equipment;
- Conditions exist during wet weather events that probably do not occur otherwise, such as flooding, lightning, and basins filling (see discussion in Section 2.0); and
- A degree of urgency and unfamiliarity exists when dealing with a wet weather event when compared to normal dry weather flows.

Because of these increased risks, performing a separate risk analysis for wet weather operations is critical to maintaining a safe working environment. Water Environment Federation's *Safety, Health, and Security in Wastewater Systems* (WEF, 2013) provides a detailed discussion on performing a risk analysis and should be referenced when performing such an analysis.

3.2 Disaster Planning

There is no substitute for good planning. Developing plans before wet weather events helps mitigate the hazards and risks involved with operations during a wet weather event. Ensuring that equipment is functioning and available for wet weather operations enhances the treatment facility's ability to process and treat wet weather flows. Placing a primary clarifier in service, even if its mechanism is not functional, can help to hydraulically pass flow (reducing flooding risks) and will provide some level of solids removal (increasing the performance of downstream processes).

Wet weather standard operating procedures (SOPs) should include preparations that can be done before the wet weather event, basic standard operating procedures to follow during an event (see Chapter 4), what to do if certain situations arise, and procedures for returning to normal operations.

These wet weather SOPs should be developed and reviewed with operations staff frequently and in advance of a wet weather event to minimize the "unfamiliarity" factor. In some instances, conducting a "dry run" is possible and can help identify actions that might seem clear when reviewing them on paper, but could be confusing when actually implementing the procedures.

Following a wet weather event, it is always a good idea to modify and update the wet weather SOPs based on lessons learned during the actual event. The reviewers should answer questions like how well were overflows prevented, what equipment failed to perform, and what equipment wasn't available that was really needed. In addition, risk mitigation should be performed to make changes at a facility in between wet weather events that will eliminate or reduce hazards. Performing this analysis also reinforces the SOPs for responding to wet weather events and helps new operators become more familiar with these procedures and how to avoid related wet weather hazards.

4.0 DISASTER MANAGEMENT AND RECOVERY

Although many wet weather management plans identify operations during the event, they must also contain procedures for coordinating with other utilities and agencies to implement the plan and to allow a safe and efficient return to normal operations after the wet weather event has passed.

Successful disaster management and recovery involves implementation of the wet weather strategy and coordination with other utilities or agencies that are involved in the same wet weather event. These include the following:

- Event monitoring and reporting to the regulatory agency,
- Coordination with local emergency operations,

- Coordination with federal agencies that provide support following a catastrophic event, and
- Collaboration with other wastewater agencies.

Every emergency plan includes preparation of a list of telephone numbers, e-mails, Web sites, and contact information. Those lists and contacts are the basis for coordination issues related to emergency response plans.

4.1 Event Monitoring and Reporting

Regulatory agencies still require monitoring reports during wet weather events as they do during normal dry weather flows. During wet weather events, operations staff may be busy addressing equipment needs; as such, it is important to make sure that adequate staff are available during wet weather to ensure that adequate data are collected. If power is lost, some sampling may have to be manually performed that might otherwise be automatically performed. Additionally, the site's National Pollutant Discharge Elimination System Industrial Multi-Sector Stormwater Permit may require stormwater sampling during a wet weather event. Manual collection of samples may pose its own safety hazards that must be addressed when developing a wet weather SOP.

Following wet weather events, additional reports may be required. This is especially true if there are sewer overflows, basement backups, or WRRF bypasses. Ensuring that the necessary data are readily available allows reports to be provided in a timely manner at a time when workloads are at a high level. Early and frequent communication with a regulatory agency should be conducted.

4.2 Coordinating with Local Emergency Operations

A wet weather management plan will identify other local agencies or emergency responders that can assist during a significant wet weather event. Local emergency responders include the following:

- Fire and paramedics,
- Law enforcement,
- Power utility, and
- Local news agency.

Inviting fire services personnel to visit a facility and evaluate the spaces and places they may have to assist with in the future is a good idea. Their assistance may be related to confined spaces, elevated towers, pits, tanks, or process areas. With wet weather events, utility management may have to meet three different shift commanders at three different times; interaction with them leads to improved communication and smoother responses. To be effective, fire prevention plans must be exercised and reviewed on a regular basis.

Law enforcement typically exists on a local level and a county/state level. Local law enforcement is typically the first to respond. If there is a wet weather event generating flooding, local law enforcement agencies will be the ones to place barricades and roadblocks to areas that pose a public risk. An important part of any emergency plan includes getting responders to the facility by coordinating the issue with local law enforcement controlling that access. Collections system personnel, as they inspect remote facilities, are in a unique position to inform local law enforcement of unsafe conditions that they may not be aware of otherwise.

For many facilities, county law enforcement is considered their local agency and the agency providing service to the facility is where coordination activities should be directed. However, it is also possible to have lift stations and other equipment in areas patrolled by county law enforcement even though the facility is in another area. Most city and county law enforcement agencies work closely together and some share common facilities and joint communication centers. Although much of the work of a wastewater utility is restricted to local facilities and systems, state law enforcement can quickly become a utility's new best friend when large weather events cause widespread catastrophic damage.

The power utility maintains electrical power to the treatment facilities. Most treatment facilities either have redundant utility power connections or have critical processes protected with a backup power generation system. During wet weather events, power outages are common and can result in the failure of some equipment as well as loss of lighting, creating an additional level of hazards. If on-site power generation exists, breakers must be opened so that there is no risk of feeding power back to the utility. When utility power is restored, appropriate methods of returning power have to be developed to prevent overloading what may be a fragile utility power system.

Local news agencies can assist with releasing public safety messages to a wide audience. Reverse 911 is also a method to reach an audience within a geographical area. These services should be used if there are chemical spills, flooding, or other warnings of conditions that may affect public health and safety (such as a boil water order) in a specific locale.

4.3 Coordination with Federal Agencies

Recovery following a catastrophic event involves first developing a clear set of priorities that allow wastewater treatment to resume as quickly as possible. This may require many changes from normal operations, from streamlining the procedures for purchasing equipment to the delegation of responsibilities to others to free up time to address high priority items.

The mission of the Federal Emergency Management Agency (FEMA) is to "support citizens and first responders to ensure that as a nation we work together to build, sustain, and improve our capability to prepare for, protect against, respond to, recover from, and mitigate all hazards". Federal assistance becomes available when the president of the United States declares a "major disaster" for the affected area at the request of a state governor. The Federal Emergency Management Agency will provide information through the media and community outreach about federal assistance and how to apply. Information on FEMA can also be accessed through the FEMA Web site at www.fema.gov.

The water/wastewater agency response network (WARN) is a mutual aid network designed to provide aid and assistance based on established agreements between water and wastewater agencies. In general, WARN members build and strengthen relationships between each other, but also define responsibilities and roles including involvement of the utility, state, and permitting agencies; state and local emergency management and response agencies; and federal or supporting professional organizations. A WARN is organized by state. The U.S. Environmental Protection Agency (U.S. EPA) and other regulatory agencies are supporting the development of the networks. Information on WARNs can be found on U.S. EPA's Web site at http://www.epa.gov/safewater/watersecurity/pubs/fs_watersecurity_warn.pdf. Contact information for local WARNs can be found at http://www.awwa.org/resources-tools/water-knowledge/emergency-preparedness/water-wastewater-agency-response-network.aspx. This network can be used to provide equipment, chemicals, and labor. Each state coordinates its network and uses a standard agreement for each utility that wants to be included in the network. This standard agreement allows the terms and conditions to be established before their use.

5.0 REFERENCE

Water Environment Federation (2013) *Safety, Health, and Security in Wastewater Systems,* 6th ed.; WEF Manual of Practice No. 1; Water Environment Federation: Alexandria, Virginia.

9

Modeling for Wet Weather

Peter A. Vanrolleghem, Ph.D., ing.; Lorenzo Benedetti; Susan Moisio; and Nalin Sahni

1.0 HOLISTIC PLANNING AND OPERATIONS—THE INTEGRATED URBAN WASTEWATER SYSTEM

1.1 Introduction

1.1.1 Benefits of Integrated Modeling

Integrated urban wastewater management presents great opportunities to minimize effects on the receiving water and associated costs. Effects to be considered include river water quality (oxygen depletion, ammonia toxicity, eutrophication, pathogens) and river morphology (erosion, fish habitat deterioration). Options like effect-based real-time control, global sewer control, construction of retention volumes and treatment facilities, and water reuse can all be better designed and evaluated by using models and data that include all subsystems involved. The problem becomes more complex, but the larger number of degrees of freedom allows for finding better solutions. Modeling of the integrated urban wastewater system (IUWS) is a powerful tool to identify synergies and to globally optimize wastewater system performance for the assessment and definition of system planning needs and for discharge permit negotiations (Benedetti, Langeveld, Comeau, Corominas, Daigger, Martin, Mikkelsen, Vezzaro, Weijers, and Vanrolleghem, 2013; Blumensaat et al., 2012).

1.1.2 Challenges of Integrated Modeling

Integration can be as simple as a time series input from the collection system model into the water resource recovery facility (WRRF) model or as complex as analyzing the sewer–WRRF–river subsystems in one integrated model. One of the main challenges for integrated modeling is the necessary data collection, especially for the parts of the system weaker in terms of model prediction capabilities and important for receiving water quality impact assessment, like sewer water quality (Langeveld, Nopens, Schilperoort, Benedetti, de Klein, Amerlinck, Weijers, 2013). Other challenges relate to the fact that the subsystem models originate from different water subdisciplines, leading to model coupling issues (Rauch et al., 2002). Refer to Figure 9.1 for a simple graphic.

1.2 Modeling of System Components

1.2.1 Urban Catchment

Phenomena responsible for runoff formation and routing to the sewer system are typically modeled by means of simple empirical equations, and can be described separately (e.g., evaporation, infiltration, wetting losses) or

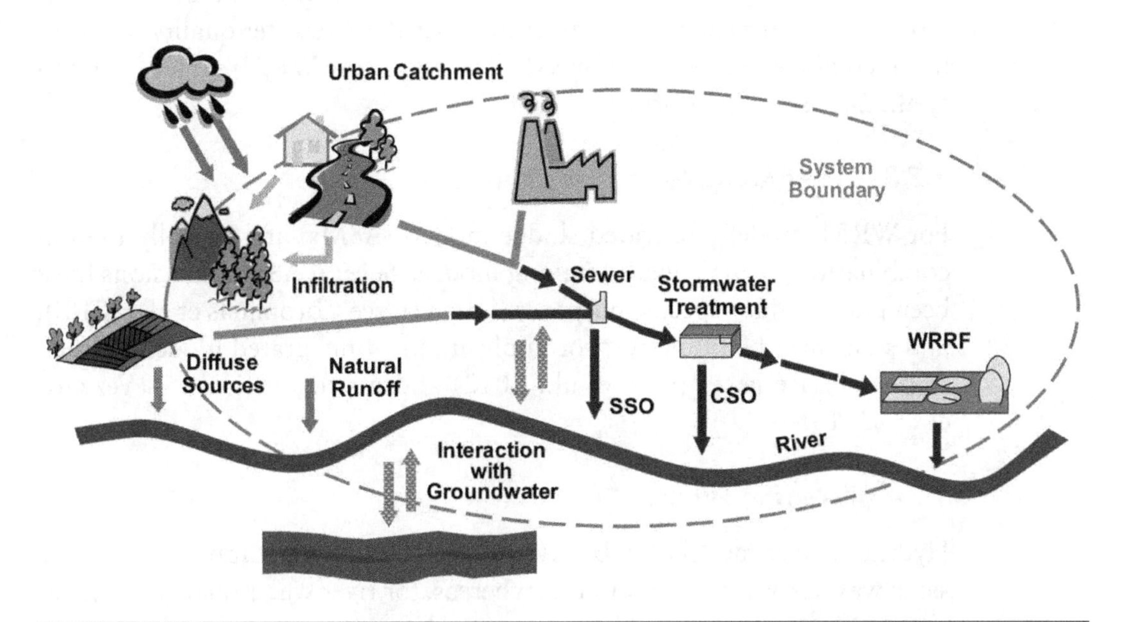

FIGURE 9.1 Simple graphic showing the components of the IUWS.

lumped by one or more reduction coefficients. Pollution generation, when considered, may include municipal and industrial dry weather flows and accumulation/wash-off of deposited pollutants (Butler and Davies, 2010). Within the catchment, source control structures also can be modeled because they alter the water and pollutant fluxes in the system (Wong et al., 2006).

1.2.2 (Simplified) Collection System

Two main approaches for collection system descriptions exist: models with full hydrodynamics (de Saint-Venant partial differential equations [PDEs]) and models with simplified so-called "hydrologic routing" (tanks-in-series [TIS] ordinary differential equations [ODEs]). In the first type of models, typically the spatial discretization is finer (a larger number of smaller catchments and pipes), which, combined with a PDE formulation, leads to considerably longer simulation times. Of course, full hydrodynamic models provide more detailed information, especially regarding water levels and velocities, but the results obtained with simplified ODE models are typically of sufficient quality for the main purpose for which they are developed, that is, predicting water flows and volumes in the context of the water quality studies focused on here. Difficulties in modeling backwater effects in the hydrologic routing approach have been dealt with recently (Vanrolleghem et al., 2009).

The reliability of water quality models for the collection system remains one of the weakest points (Bertrand-Krajewski, 2007). Several process models have been published, but their applicability is typically limited to the local conditions for which they were developed. Alternatives to process models are empirical models based on detailed wastewater quality monitoring at combined sewer overflows (CSOs) and at the WRRF's inlet (Mourad et al., 2006; Schilperoort et al., 2012).

1.2.3 Water Resource Recovery Facility

For WRRF models, activated sludge models (ASMs) are typically used in combination with TIS-based mixing models. Several ASM extensions have been published for specific purposes (for a list, see Corominas et al. [2010]), and some may be of interest for applications of integrated modeling, such as greenhouse gases (Guo et al., 2012) and microconstituents (Vezzaro et al., 2014).

1.2.4 Receiving Stream

Hydraulic river models can be divided in PDE and ODE approaches in the same way as for the sewer models; whereas, for river water quality, the main advancements toward model integration have happened already through the work of Reichert et al. (2001) with the harmonization of state variables in WRRF models and the river water quality model (Rauch et al., 2002).

1.2.5 Interfaces between Subsystems

One of the challenges in model integration is the linking between the different subsystem models, which typically have different sets of state variables. The simplest approach is to fractionate or aggregate analogous state variables, developing one model connector for each couple of models (Benedetti et al., 2004). This approach has evolved into a formalized method that ensures closed elemental mass and charge balances (Vanrolleghem, Rosen, Zaher, Copp, Benedetti, Ayesa, and Jeppsson, 2005). Another option that would be available, but has so far been applied only to linking unit models within the WRRF fence, is to develop a model that can be applied to all system units with a single superset of components (thereby alleviating the need for coupling) (Grau et al., 2007).

Subsystem models built in different subsystem-specific software can be coupled by passing properly formatted data files between the programs or by software integration tools such as OpenMI (www.openmi.org). In the case of "simplified" models (e.g., with only ODEs), the possibility to implement all of them in the same modeling software also is available.

1.2.6 Real-Time Control

Real-time control can be included in integrated models in any part of the system, locally (WRRF, collection system, river) or globally, that is, with information flow between the subsystems such as ammonia and dissolved oxygen sensors in the river driving flows in the collection system and aeration in the WRRF (Corominas et al., 2013; Langeveld, Benedetti, de Klein, Nopens, Amerlinck, van Nieuwenhuijzen, Flameling, van Zanten, and Weijers, 2013; Vanrolleghem, Benedetti, and Meirlaen, 2005).

1.3 Generation and Evaluation of Options

1.3.1 Simulation Period: Event-Based Versus Long-Term Simulation

With the developments in computational efficiency (both hardware and software), the possible uses of integrated modeling have largely expanded. It is currently possible to run long-term integrated simulations (e.g., 1 year). Event simulations may be used to test the validity and behavior of models under specific circumstances; however, to evaluate the response of such a complex system, it is necessary to use long-term simulations because of the interactions between storage volumes, their emptying time, hydraulic and treatment capacity at the WRRF, assimilation capacity of the river, and other factors. Indeed, short-term events can have long-term effects, for example, sediment oxygen demand or eutrophication induced by nutrient-rich CSO events (see Figure 9.2 for an example).

1.3.2 Key Performance Indicators, Time Series Analysis

Critical to any modeling study is defining key performance indicators (KPIs) on which the system will be evaluated. In an integrated modeling study, they can be the chemical and/or ecological quality of the receiving stream, the number of sewer overflow events, operating costs, the level of process protection at the WRRF, or many other factors. Mostly, such KPIs involve a statistical analysis of simulated time series, as criteria may be defined on the basis of maximum month values, return periods, and percentage of time in compliance, like in the *Urban Pollution Management Manual* (http://www.fwr.org/UPM3). After the KPIs are set, the required model is defined in terms of its boundaries; its spatial, temporal, and process detail; and the simulation period.

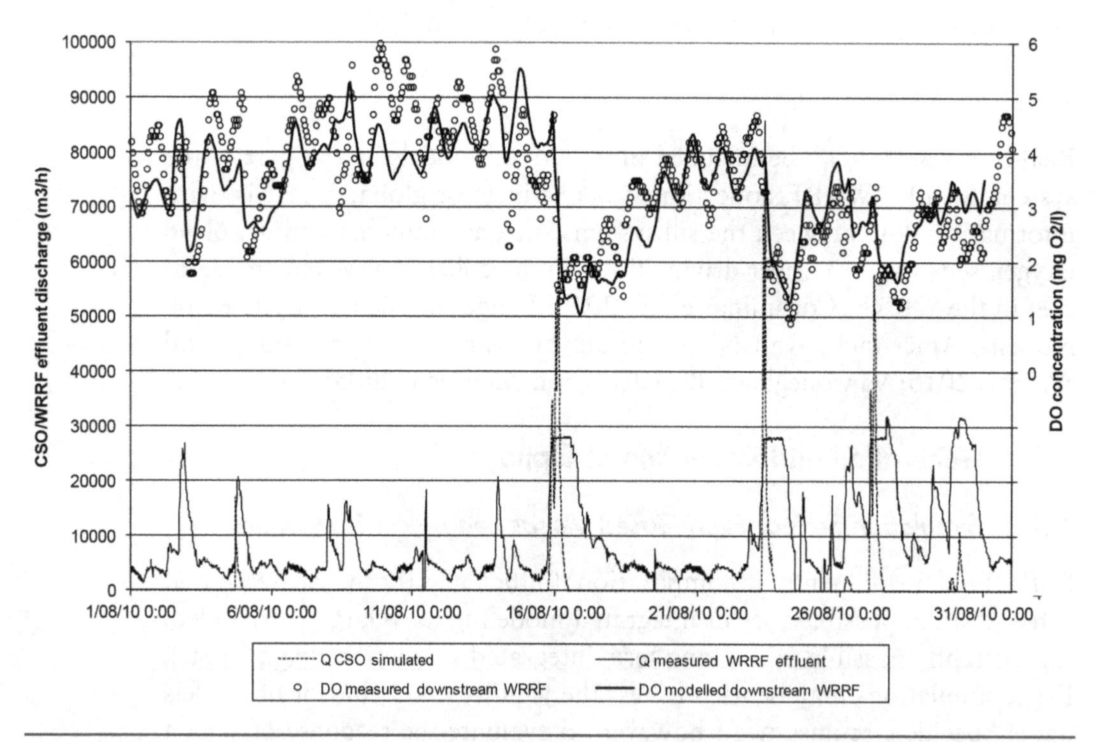

FIGURE 9.2 Performance of integrated model (dissolved oxygen in a river section downstream of the WRRF and the CSOs) for 1 month of simulation with typical wet weather events leading to dissolved oxygen depletion (DO = dissolved oxygen) (from Langeveld et al. [2013]).

1.4 Examples of Case Studies

In Denmark, models are used by the water utilities to evaluate and improve the performance of real-time control strategies, which include weather-radar measurements and discharging to the least sensitive receiving waterbody (Poulsen et al., 2013). In the Netherlands, modeling has been used to derive an optimal and robust set of actions in the integrated system to meet the regulation requirements at the least costs (Benedetti, Langeveld, de Jonge, de Klein, Flameling, Nopens, van Nieuwenhuijzen, van Zanten, and Weijers, 2013).

The Metropolitan Sewer District of Greater Cincinnati—Sustainable Watershed Evaluation Planning Process is an integrated planning process using KPIs. It is following the lead of the U.S. Environmental Protection Agency's (U.S. EPA's) Integrated Municipal Stormwater and Wastewater Planning Approach Framework (U.S. EPA, 2012) and focusing its efforts on meeting the objectives of the Clean Water Act (CWA) and meeting requirements of its federal mandates using this integrated planning framework. Based on these overarching goals, examples of watershed-specific goals and objectives include eliminating all sanitary sewer overflow (SSO) events, reducing the volume of CSOs, addressing local flooding issues, and providing community benefits. Examples of KPIs include compliance with consent decrees, progress toward CWA standards, improvement of workforce skills, an increase in viable housing options, or enhancement of community amenities and characteristics.

2.0 MODELING THE INFLUENT CHARACTERISTICS—A COLLECTION SYSTEM PERSPECTIVE

The inputs necessary for modeling the influent characteristics of the collection system are dependent on the end use of the model.

2.1 Flow Characterization

2.1.1 Flow Generation

2.1.1.1 Combined Systems: Rainfall/Runoff/Groundwater Infiltration/Snowmelt

The modeled flows generated in a combined sewer subcatchment are a function of the subcatchment characteristics. Characteristics such as the area, land use, ground slope, ground cover (roughness coefficient), land use (pervious and impervious areas), and the soil parameters affect the amount of flow generated and the flow constituents. The demographic distribution along with industrial development also affects the amount of flow generated in the combined system.

Seasonal conditions such as snow accumulation and snowmelt are also factors that affect the inflow and infiltration modeled in a combined subcatchment. Snowmelt is modeled in a combined system model if the integrated system is affected by seasonal snowmelt.

2.1.1.2 Separate Systems: Rainfall-Dependent Inflow and Infiltration

Inflow and infiltration is modeled in separate sewer systems because the sewer system response to this excess flow can vary dramatically. Inflow is the stormwater that enters into the sewer system as a direct connection. An example of a direct connection would be a downspout or a catch basin. Infiltration is stormwater that enters the sewer system by infiltrating through the soil. The modeler should understand the collection sewer system response that represents the full range of flow conditions. Some infiltration in a sanitary sewer system is expected; however, rainfall-dependent inflow and infiltration (RDII) is excessive inflow that enters a collection system during rainfall. This should be differentiated from seasonal groundwater infiltration. Because RDII enters the sanitary sewer system through defects or cross connections, this model input should be measured and not calculated. Data from field studies can be used to estimate the amount of RDII in a sanitary system, but flow monitoring data are recommended. When modeling a wet sanitary sewer system continuously, seasonally varied RDII factors must be determined. The modeler must determine the dry weather flow (base wastewater flow plus groundwater infiltration) and differentiate this flow from RDII. Refer to Figure 9.3 for an example of the components of flow.

2.1.2 Flow Conveyance/Storage: Operations with Special Attention to Storage

Collection system elements such as real-time control facilities, real-time control strategies, and storage and conveyance elements should be a part of the collection system model. These elements affect the peak flow response and volume of flow at the WRRF and at CSOs and SSOs.

2.1.3 Modeling Climate Change

Modeling scenarios reflecting the potential effects of climate change should be a part of an integrated modeling strategy. Understanding the effect of climate challenge to the collection system includes elements such as the following:

- Drought:
  - Reduced groundwater recharge,
  - Lower lake and reservoir levels,

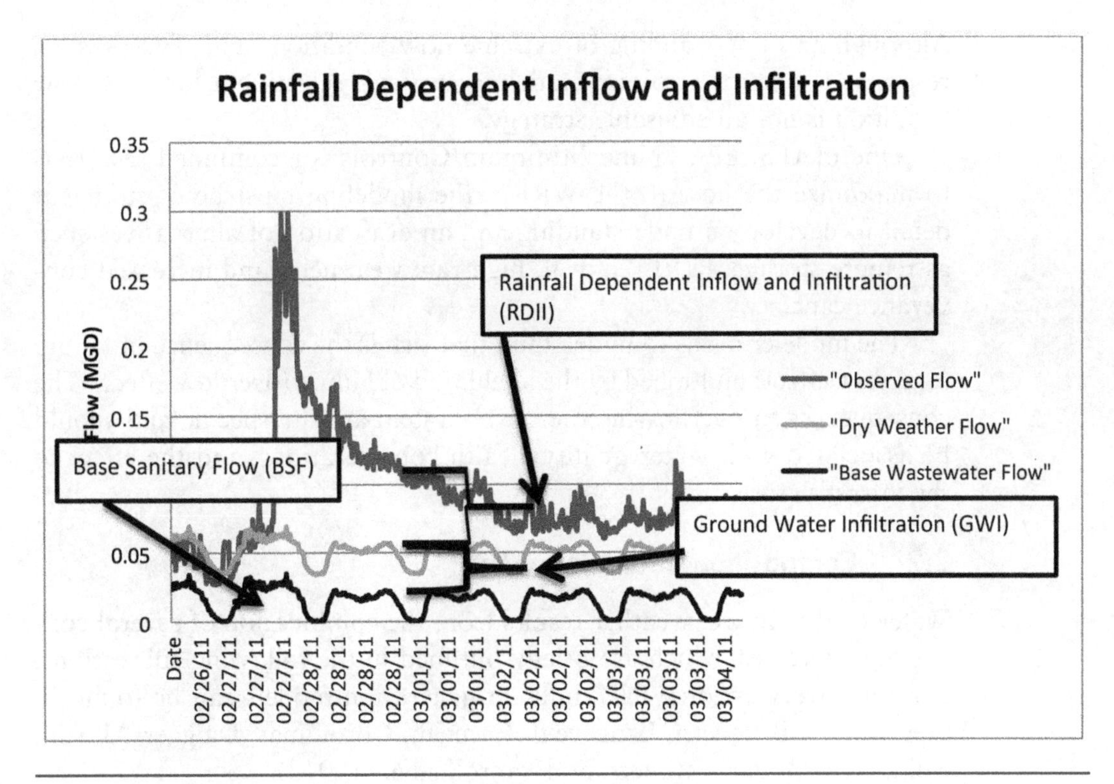

FIGURE 9.3 Hydrograph disaggregation showing flow components.

  - Changes in seasonal runoff and loss of snowpack, and
  - Fire hazard risk to facilities and watersheds.
- Water quality degradation:
  - Low flow conditions;
  - Altered water quality (higher temperature, algal blooms, increased runoff);
  - Saltwater intrusion into aquifers from sea level rise; and
  - Altered surface water quality.
- Floods:
  - Increased frequency of high flow events and flooding and
  - Flooding from coastal storm surges and sea level rise.

These changed conditions need to be considered to model a robust future condition.

2.1.4 Risk/Regulatory Effects

The collection system model should take into account the risk of a particular peak flow response occurring and the regulatory effects of this response.

Although an understanding of extreme flow conditions and sewer system response is part of a robust modeling strategy, selecting a low-risk flow condition is not an advisable strategy.

One of U.S. EPA's Nine Minimum Controls for combined sewers is to maximize the flow to the WRRF; the modeling must be of sufficient detail to develop an understanding and an evaluation of alternatives such as remote storage, RDII removal, high rate treatment, and increased conveyance capacity.

The modeler needs to understand that risk is the consequence of failure (overflow effect) multiplied by the likelihood of failure (overflow effect). The consequences of overflowing the sewer system are site-specific and should be associated with water quality, health considerations, and the effect to the infrastructure.

2.2 Composition Characterization

Water quality in wet weather results from the combination of several concurring processes, which are briefly outlined in the following subsections. A recent overview of WRRF influent generation models can be found in literature by Benedetti, Langeveld, Comeau, Corominas, Daigger, Martin, Mikkelsen, Vezzaro, Weijers, and Vanrolleghem (2013).

2.2.1 Runoff Pollution Composition in Combined Sewers

Constituents entering the collection system from urban surfaces are partly associated with the solids mobilized by runoff from roads, parking lots, and roofs and partly from the solubilization of substances present in or on surfaces. Next to organics, they may also be responsible for a significant fraction of the WRRF load regarding heavy metals and organic xenobiotics. In cold climates, deicing salts entering a combined system constitute a further challenge. Such phenomena can be modeled either with a fixed concentration or load per event or by using dynamic functions that quantify the accumulation of substances on the catchment surface and their subsequent wash-off (Butler and Davies, 2010; Sansalone and Kim, 2008).

2.2.2 First Flush (Sediment Resuspension Plus Dry Weather Plug Flush-Out)

The increase in flowrate at the beginning of the rain event has the effect of "pushing", like a plug, the dry weather constituents present in the collection system, with a short but intense increase in load to the WRRF, which may induce effluent ammonium peaks. In case the antecedent dry period allowed for sediment deposition and the flowrate generates sufficient shear stress for resuspension, the so called "first flush" effect brings a substantial solids peak load to the facility, with consequent possible settling problems in the WRRF. To model those peak loads and the following period with diluted influent, empirical models are typically used that correlate flowrate with loads or concentrations. Another important difference between dry and wet weather concerns the influent chemical oxygen demand (COD) composition, which shows lower soluble and readily biodegradable fractions resulting from the shorter retention time and lower temperature in the sewer, which reduces biological nutrient removal. This can be reproduced by adopting different COD fractionation parameters in different weather conditions (Bixio et al., 2001).

2.2.3 Sediment Inventory

To properly model the dynamics of the particulates entering a treatment facility, it is important to properly describe the evolution of the sediment inventory in a sewer system. Sedimentation and resuspension are the processes that determine the accumulation of particulate mass in the sewer system. Although a number of attempts have been made to model this behavior, the quality of the water quality predictions remains fairly low (Bertrand-Krajewski, 2007; Butler and Davies, 2010). Given the important research efforts in this area, it can be expected that improved models become available in the not-too-distant future.

2.2.4 Retention Tank Effects on Wastewater Quality

Retention tanks installed in the sewer system to reduce CSOs also have an effect on the wastewater composition. Sedimentation and resuspension affect the particulate concentration profiles that arrive at the WRRF after the retained water is released again. Sometimes, particular ways of cleaning and emptying the tanks induce important peak loads of particulates to the WRRF (Maruéjouls et al., 2013).

2.2.5 Temperature Variations Resulting from Rain, Snowmelt, or Summer Storms

Temperature decreases associated with rain and especially snowmelt have significant detrimental effects on process performance (nitrification, settling) at the WRRF (Plósz et al., 2009). Such effects will become more pronounced because of climate change and should be considered in a temperature or heat model of the WRRF.

2.2.6 Oxygen Content

After the initial phase of a wet weather event, the influent becomes diluted and may start to contain significant dissolved oxygen (>1 g O_2/m^3). Such conditions will also deteriorate the performance of anaerobic and/or anoxic processes placed at the head of the WRRFs, resulting in lower nutrient removal rates.

3.0 MODELING WET WEATHER AT THE WATER RESOURCE RECOVERY FACILITY

3.1 Preliminary Treatment

In terms of modeling WRRFs, little emphasis has been put on the preliminary treatment steps. Mostly, the model limits itself to a removal percentage of particulates. However, under wet weather conditions, these preliminary treatment steps can become critical systems because they may, for instance, be subject to clogging (screens) or considerably reduced efficiency (grit chambers). Proper accounting for the induced water quality alterations should be pursued and, for the time being, modeled using empirical relationships.

3.2 Primary Treatment

Although many WRRF models start at the primary effluent, attention is increasing to include primary treatment models in view of the importance of maximizing organics to be sent to anaerobic digestion for energy recovery (biogas, electricity) through, for example, chemical addition. Simple (ideal point settler) primary clarifier models suffice mostly for simulations under dry weather conditions, but the reduced efficiency of primary clarification must be accounted for in wet weather conditions. A few modeling options exist to describe the effect of the increased hydraulic and organic loading on separation performance.

3.2.1 Models with Enhanced Predictive Power for Wet Weather Flows

The benchmark simulation community (Jeppsson et al., 2007) selected the Otterpohl model for its simplicity and ability to describe reduced clarification efficiency under wet weather conditions. Using the layer approach to represent suspended solids profiles in the clarifier allows representing accumulation of suspended solids in the clarifier and their scouring as shear increases with wet weather flow (Lessard and Beck, 1999). A number of similar alternatives are also available.

3.2.2 Chemically Enhanced Primary Treatment Models

One of the ways to handle increased wet weather loadings is the use of chemicals to enhance primary treatment. Their effect can simply be modeled empirically by increasing the removal efficiency as a function of the applied dose (Tik et al., 2013), but newer models that consider coagulation and flocculation explicitly will become available in the coming years.

3.3 Biological Treatment

3.3.1 Influent Wastewater Characterization for Model Use

Although the overall changes in wastewater composition under wet weather have been discussed in section 2.2 (e.g., first flush), it is important to realize that the fractionation of, for instance, suspended solids changes drastically from dry weather composition. For instance, an increased fraction of inert inorganics (sand) can be expected to occur. In addition, material that is resuspended may have undergone degradation as it resided in the collection system and a less biodegradable fractionation of suspended solids may be expected. So far, few studies have been conducted to perform wastewater fractionation of wet weather samples.

3.3.2 Modeling Mixing Behavior under Wet Weather Hydraulics

It is well known that flow conditions (e.g., expressed as retention time) affect the flow and mixing regime in bioreactors.

3.3.2.1 Tanks-in-Series

To illustrate the effect of the different flow conditions, one can turn to the following simple equation describing the number of TIS that are necessary to describe mixing in an activated sludge reactor (Shaw and Johnson, 2013):

$$N = 7.4\frac{L}{WH}Q_{in} \quad (9.1)$$

This equation directly shows that an increased flow, Q_{in}, through a reactor with length, L, width, W, and height, H, will lead to an increased number of TIS, that is, behavior that is more plug flow is expected. Although this feels counterintuitive as the increased hydraulic load leads to increased turbulence, one must consider that there is less (retention) time for this turbulence to have an effect and the overall result is reduced mixing.

3.3.2.2 Computational Fluid Dynamics

Behavior of bioreactors is increasingly being studied using computational fluid dynamics. These modeling methods are particularly suited to study the effect of changes in hydraulic load on mixing behavior in large-scale activated sludge reactors (Wicks and Wicklein, 2012).

3.3.3 Modeling Aeration Efficiency under Wet Weather Conditions

It is well known that process conditions affect aeration efficiency. Wet weather induces composition changes that affect aeration and the temperature changes that come with wet weather may also affect mass transfer efficiency. These effects can be modeled by appropriately adjusting the alpha and beta factors of the mass transfer relationship.

3.3.4 Secondary Clarifier Models with Enhanced Predictive Power under Wet Weather Flows

The significant shift in the sludge inventory with a displacement of sludge from the bioreactors to the secondary clarifiers is one of the most important effects of wet weather conditions and has thus received ample attention. In principle, a model describing sludge accumulation well is the minimum requirement. The industry standard Stenstrom-Vitasovic-Takacs model (Takacs et al., 1991) is considered adequate for this task, although it has recently been challenged for its inability to adequately describe compression phenomena (Bürger et al., 2012).

3.3.5 Modeling Approaches for Real-Time Control Systems for Wet Weather

Adequate modeling of activated sludge facilities is sometimes hampered by the difficulties encountered to properly describe the control systems with which an existing WRRF is equipped. Often, rule-based controllers are implemented that are poorly documented, which leads to large deviations between real and modeled WRRFs. Control systems often have special rules for the operation under wet weather flow and these must be reproduced to represent the system.

Furthermore, some control actions lead to particular behavior at the facility. Although ratio control of the return activated sludge flow or step feed are easy to deal with from a modeling perspective, an approach such as aeration rank settling requires additional modeling effort (Benedetti et al., 2011). In this system, the bioreactors become sedimentation tanks to reduce the sludge load on the secondary clarifiers similar to step feed, but its modeling requires the conversion of a single tank bioreactor model to a two tanks in parallel model with different concentrations plus a settler model.

3.4 Physical and Chemical Treatment

Modeling physical and chemical treatment unit processes that are part of water WRRFs is currently receiving little attention from the modeling community. In view of the increasing attention to disk filters, advanced oxidation, membrane filtration, and so on, it can be expected that models will be introduced and the extensions to wet weather flow conditions developed.

3.5 Disinfection

The efficiency of disinfection through physical (e.g., UV) or chemical means (e.g., chlorination) is affected by wet weather flow conditions because, for example, more (particulate) organics leave with the effluent. Unfortunately, modeling of these systems is not well developed and the modeling of these effects has hardly been touched on. Empirical modifications to disinfection efficiency parameters can be provided to mimic the effect.

3.6 Residuals Processing and Management

The increased influent loading with particulates may create a sudden excess sludge production that may overload the residuals processing equipment. Having models that help describe the performance of these unit processes under these overloading conditions is essential in view of their proper design and operation. Typically, the models used are applied and empirical modifications are made to deal with the overloads. The performance of the models under these conditions should be carefully evaluated.

4.0 REFERENCES

Benedetti, L.; Langeveld, J. G.; Comeau, A.; Corominas, L.; Daigger, G. T.; Martin, C.; Mikkelsen, P.-S.; Vezzaro, L.; Weijers, S.; Vanrolleghem, P. A. (2013) Modelling and Monitoring of Integrated Urban Wastewater Systems: Review on Status and Perspectives. *Water Sci. Technol.,* **68** (6), 1203–1215.

Benedetti, L.; Langeveld, J. G.; de Jonge, J.; de Klein, J. J. M.; Flameling, T.; Nopens, I.; van Nieuwenhuijzen, A.; van Zanten, O.; Weijers, S. (2013) Cost-Effective Solutions for Water Quality Improvement in the Dommel River Supported by Sewer-WWTP-River Integrated Modelling. *Water Sci. Technol.,* **68** (5), 965–973.

Benedetti, L.; Meirlaen, J.; Vanrolleghem, P. A. (2004) Model Connectors for Integrated Simulations of Urban Wastewater Systems. In *Sewer Networks and Processes within Urban Water Systems*; Bertrand-Krajewski, J.-L., Almeida, M., Matos, J., Abdul-Talib, S., Eds.; IWA Publishing: London, U.K.; pp 13–21.

Benedetti, L.; Nyerup Nielsen, C.; Thirsing, C. (2011) Modelling for Integrated Sewer-WWTP Operation with ATS in Copenhagen. *Proceed-*

ings of the 12th Nordic Wastewater Conference; Helsinki, Finland, Nov 14–16.

Bertrand-Krajewski, J.-L. (2007) Stormwater Pollutant Loads Modelling: Epistemological Aspects and Case Studies on the Influence of Field Data Sets on Calibration and Verification. *Water Sci. Technol.,* **55** (4), 1–17

Bixio, D.; van Hauwermeiren, P.; Thoeye, C.; Ockier, P. (2001) Impact of Cold and Dilute Sewage on Pre-Fermentation —A Case Study. *Water Sci. Technol.,* **43** (11), 109–117.

Blumensaat, F.; Staufer, P.; Heusch, S.; Reußner, F.; Schütze, M.; Seiffert, S.; Gruber, G.; Zawilski, M.; Rieckermann, J. (2012) Water Quality-Based Assessment of Urban Drainage Impacts in Europe—Where Do We Stand Today? *Water Sci. Technol.,* **66** (2), 304–313.

Bürger, R.; Diehl, S.; Farås, S.; Nopens, I. (2012) On Reliable and Unreliable Numerical Methods for the Simulation of Secondary Settling Tanks in Wastewater Treatment. *Comp. Chem. Eng.,* **41**, 93–105.

Butler, D.; Davies, J. W. (2010) *Urban Drainage,* 3rd ed.; Spon Press: New York; p 632.

Corominas, Ll.; Acuña, V.; Ginebreda, A.; Poch, M. (2013) Integration of Freshwater Environmental Policies and Wastewater Treatment Plant Management. *Sci. Total Environ.,* **445–446**, 185–191.

Corominas, Ll.; Rieger, L.; Takács, I.; Ekama, G.; Hauduc, H.; Vanrolleghem, P. A.; Oehmen, A.; Gernaey, K. V.; van Loosdrecht, M. C. M.; Comeau, Y. (2010) New Framework for Standardized Notation in Wastewater Treatment Modeling. *Water Sci. Technol.,* **61** (4), 841–857.

Grau, P.; de Gracia, M.; Vanrolleghem, P. A.; Ayesa, E. (2007) A New Plant-Wide Modelling Methodology for WWTPs. *Water Res.,* **41** (19), 4357–4372.

Guo, L.; Porro, J.; Sharma, K.; Amerlinck, Y.; Benedetti, L.; Nopens, I.; Shaw, A.; Vanrolleghem, P. A.; Van Hulle, S. W. H.; Yuan, Z. (2012) Towards a Benchmarking Tool for Minimizing Wastewater Utility Greenhouse Gas Footprints. *Water Sci. Technol.,* **66** (11), 2483–2495.

Jeppsson, U.; Pons, M.-N.; Nopens, I.; Alex, J.; Copp, J. B.; Gernaey, K. V.; Rosen, C.; Steyer, J.-P.; Vanrolleghem, P. A. (2007) Benchmark Simulation Model No 2: General Protocol and Exploratory Case Studies. *Water Sci. Technol.,* **56** (8), 67–78.

Langeveld, J. G.; Benedetti, L.; de Klein, J. J. M.; Nopens, I.; Amerlinck, Y., van Nieuwenhuijzen, A.; Flameling, T.; van Zanten, O.; Weijers, S. (2013) Impact-Based Integrated Real-Time Control for Improvement of the Dommel River Water Quality. *Urban Water J.,* **10** (5), 312–329.

Langeveld, J.; Nopens, I.; Schilperoort, R.; Benedetti, L.; de Klein, J.; Amerlinck, Y.; Weijers, S. (2013) On Data Requirements for Calibration of Integrated Models for Urban Water Systems. *Water Sci. Technol.*, **68** (3), 728–736.

Maruéjouls, T.; Lessard, P.; Pelletier, G.; Vanrolleghem, P. A. (2013) Characterisation of Retention Tank Water Quality: Particle Settling Velocity Distribution and Retention Time. *Water Qual. Res. J. Can.*, **48** (4), 321–332.

Mourad, M.; Bertrand-Krajewski, J.-L.; Chebbo, G. (2006) Design of a Retention Tank: Comparison of Stormwater Quality Models with Various Levels of Complexity. *Water Sci. Technol.*, **54** (6–7), 231–238.

Plósz, B.; Liltved, H.; Ratnaweera, H. (2009) Climate Change Impacts on Activated Sludge Wastewater Treatment: A Case Study from Norway. *Water Sci. Technol.*, **60** (2), 533–541.

Poulsen, T. S.; Öennert, T. B.; Rasmussen, M. R.; Pedersen, J. S.; Thirsing, C. (2013) Weather Radars Used in Online Control of Wastewater Systems. *Proceedings of the 11th IWA Conference on Instrumentation, Control and Automation (ICA2013);* Narbonne, France, Sept 18–20.

Rauch, W.; Bertrand-Krajewski, J.-L.; Krebs, P.; Mark, O.; Schilling, W.; Schütze, M.; Vanrolleghem, P. A. (2002) Deterministic Modelling of Integrated Urban Drainage Systems. *Water Sci. Technol.*, **45** (3), 81–94.

Reichert, P.; Borchardt, D.; Henze, M.; Rauch, W.; Shanahan, P.; Somlyody, L.; Vanrolleghem, P. A. (2001) *River Water Quality Model No. 1 (RWQM1)*; IWA Scientific and Technical Report No. 12; IWA Publishing: London, U.K.

Schilperoort, R. P. S.; Dirksen, J.; Langeveld, J. G.; Clemens, F. H. L. R. (2012) Assessing Characteristic Time and Space Scales of In-Sewer Processes by Analysis of One Year of Continuous In-Sewer Monitoring Data. *Water Sci. Technol.*, **66** (8), 1614–1620.

Sansalone, J. J.; Kim, J. Y. (2008) Transport of Particulate Matter Fractions in Urban Source Area Pavement Surface Runoff. *J. Environ. Qual.*, **37** (5), 1883–1893.

Shaw A. R.; Johnson B. R., Eds. (2013) *Wastewater Treatment Process Modeling;* Manual of Practice No. 31; Water Environment Federation: Alexandria, Virginia.

Tik, S.; Langlois, S.; Vanrolleghem, P. A. (2013) Establishment of Control Strategies for Chemically Enhanced Primary Treatment Based on Online Turbidity Data. *Proceedings of the 11th IWA Conference on Instrumentation, Control and Automation (ICA2013)*; Narbonne, France, Sept 18–20.

U.S. Environmental Protection Agency (2012) Integrated Municipal Stormwater and Wastewater Planning Approach Framework. http://www.wefnet.

org/CleanWaterActIPF/Integrated%20Planning%20%20Framework%20 (06.05.12).pdf (accessed May 2014).

Vanrolleghem, P. A.; Benedetti, L.; Meirlaen, J. (2005) Modelling and Real-Time Control of the Integrated Urban Wastewater System. *Environ. Model. Soft.,* 20 (4), 427–442.

Vanrolleghem, P. A.; Kamradt, B.; Solvi, A.-M.; Muschalla, D. (2009) Making the Best of Two Hydrological Flow Routing Models: Nonlinear Outflow-Volume Relationships and Backwater Effects Model. *Proceedings of 8th International Conference on Urban Drainage Modelling (8UDM)* [CD-ROM]; Tokyo, Japan, Sept 7–11.

Vanrolleghem, P. A.; Rosen, C.; Zaher, U.; Copp, J.; Benedetti, L.; Ayesa, E.; Jeppsson, U. (2005b) Continuity-Based Interfacing of Models for Wastewater Systems Described by Petersen Matrices. *Water Sci. Technol.,* 52 (1–2), 493–500.

Vezzaro, L.; Benedetti, L.; Gevaert, V.; De Keyser, W.; Verdonck, F.; De Baets, B.; Nopens, I.; Cloutier, F.; Vanrolleghem, P. A.; Mikkelsen, P.-S. (2014) IUWS_MP: A Model Library for Dynamic Transport and Fate of Micropollutants in Integrated Urban Wastewater and Stormwater Systems. *Environ. Model. Soft.,* 53, 98–111.

Wicks, J. D.; Wicklein, E. (2012) Overview of Current Applications of CFD in WWTP. *Computational Fluid Dynamics Workshop at the 3rd IWA/WEF Wastewater Treatment Modelling Seminar (WWTmod2012)*; Mont-Sainte-Anne, Québec, Canada, Feb 26–28.

Wong, T. H. F.; Fletcher, T. D.; Duncan, H. P.; Jenkins, G. A. (2006) Modelling Urban Stormwater Treatment—A Unified Approach. *Ecol. Eng.,* 27 (1), 58–70.

5.0 SUGGESTED READINGS

American Public Works Association; American Society of Civil Engineers; National Association of Clean Water Agencies; Water Environment Federation (2010) *Core Attributes of Effectively Managed Wastewater Collection Systems;* American Public Works Association: Washington, D.C.; American Society of Civil Engineers: Reston, Virginia; National Association of Clean Water Agencies: Washington, D.C.; Water Environment Federation: Alexandria, Virginia.

National Association of Clean Water Agencies (2009) Confronting Climate Change: An Early Analysis of Water and Wastewater Adaptation

Costs. http://www.nacwa.org/images/stories/public/2009–1028ccreport.pdf (accessed June 2011).

Water Environment Federation (2010) *Wastewater Collection Systems Management,* 6th ed.; Water Environment Federation: Alexandria, Virginia.

Water Environment Federation (2011) *Prevention and Control of Sewer System Overflows,* 3rd ed.; Water Environment Federation: Alexandria, Virginia.

U.S. Environmental Protection Agency (1995) *Combined Sewer Overflows: Guidance for Nine Minimum Controls;* U.S. Environmental Protection Agency: Washington, D.C.

U.S. Environmental Protection Agency (2007) *Computer Tools for Sanitary Sewer System Capacity Analysis and Planning;* U.S. Environmental Protection Agency: Washington, D.C.

U.S. Environmental Protection Agency (2012) *Climate Adaptation Strategies Guide for Water Utilities,* January. http://water.epa.gov/infrastructure/watersecurity/climate/upload/epa817k11003.pdf (accessed May 2014).

U.S. Environmental Protection Agency (2014) *Climate Adaptation Strategies Guide,* and its companion tool, CREAT. http://water.epa.gov/infrastructure/watersecurity/climate/ (accessed May 2014).

Water Utility Climate Alliance (2010) *Decision Support Planning Methods: Incorporating Climate Change Uncertainties into Water Planning.* http://www.wucaonline.org/assets/pdf/pubs_whitepaper_012110.pdf (accessed Sept 2011).

10

On-Site Storage/Flow Equalization

Thomas A. Lyon, P.E.

1.0 INTRODUCTION

This chapter describes the use of on-site storage or flow equalization facilities at a water resource recovery facility (WRRF) site to dampen peak flows and wet weather volumes that could upset or wash out individual treatment processes. The emphasis is on equalizing wet weather flows, as opposed to equalizing diurnal fluctuations in dry weather flows or waste loadings.

The most important function of the flow equalization facilities is to protect the integrity of downstream treatment processes by limiting flowrates and process loadings to acceptable levels, ensuring that discharge permit conditions will be met. Before flow equalization facilities can be properly sized and situated in the process stream, two important analyses must be made: (1) existing and projected influent flow conditions and (2) peak sustainable capacities of existing unit processes.

On-site flow equalization facilities are used in response to wet weather conditions, although they may be used more often to equalize diurnal fluctuations in dry weather flows or waste loadings. Depending on the nature of the collection system and the length of the wet weather period, the facilities could be used for an hour or two or for several days. Several factors are unique to intermittent operation of wastewater facilities, and particularly to flow equalization facilities, including the following:

- Basins, junction or flow-splitting structures, and channels or pipelines that go from empty to partially full to possibly completely full and then back to empty;
- Solids deposition and odor potential;
- Freeze potential;
- Ultraviolet sunlight exposure for equipment;
- Widely varying operating levels (water surfaces);
- Odor potential from fall over weirs and similar turbulent flow conditions or from long storage periods;
- Increased housekeeping and cleaning requirements;
- Increased manpower requirements and overtime; and
- Overflow protection from wet weather events in excess of design conditions.

2.0 LOCATION

2.1 Location in the Treatment Process

Flow equalization can be upstream of preliminary treatment, downstream of preliminary treatment, or downstream of primary treatment and may incorporate use of existing, unused basins as a first stage. Typical arrangements for flow equalization facilities are illustrated in Figure 10.1, a flow equalization schematic. Note the requirements for extra capacity that are typically associated with each arrangement.

2.2 Hydraulic Considerations

Gravity influent is often preferred to provide more fail-safe treatment process protection, minimize the size of pumps required, lessen the need for

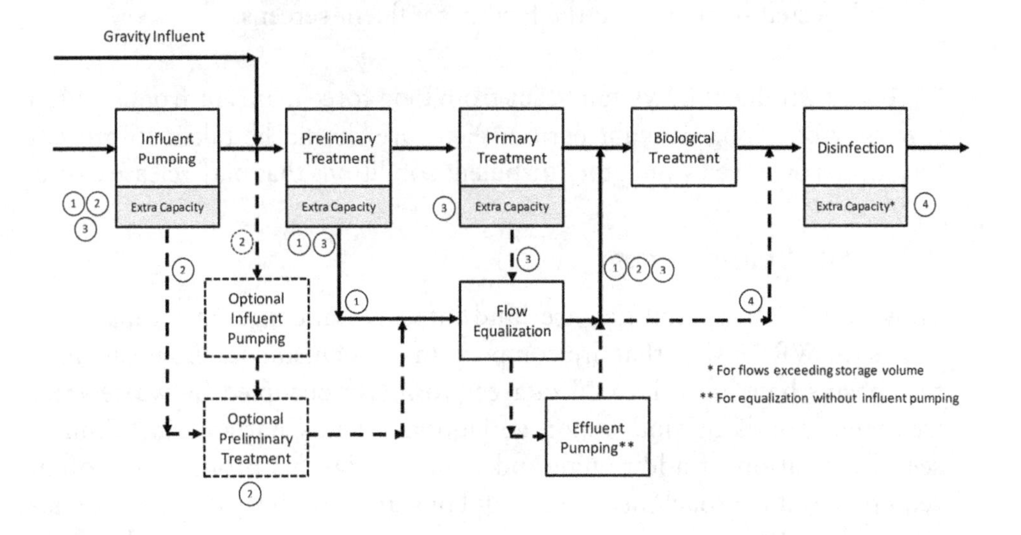

FIGURE 10.1 Typical flow equalization arrangements at a WRRF.

expensive variable-speed pumps, and minimize standby power requirements. In addition, pumped return of equalized flows to treatment processes requires smaller pumps than those required for influent pumping and can be more easily controlled.

The following types of flow-splitting designs can minimize pretreatment needs for flow equalization by directing grit and screenings to the existing preliminary treatment facilities (see Chapter 6 for hydraulic design details):

- Flow over a diversion weir in lieu of through a throttled gate;
- Overflow from the dry weather wet-well compartment to a new wet weather wet-well compartment at an expanded influent pumping station; and
- Use of a combined sewer overflow (CSO) type screen in conjunction with a side-channel diversion weir; channel velocities must be maintained in the direction of the principal screening facilities to transport deflected screenings to the facility's influent screens.

If septic influent flows can occur from long force mains or from scoured sewers after a long drought period, then care should be taken to prevent long drops over weirs and other turbulent conditions that may release odors.

2.3 Site Considerations

Flow equalization facilities are land-intensive and present siting challenges for WRRF sites that are compact. In many instances, flow equalization basins have been located in areas formerly occupied by wastewater treatment ponds or sludge storage lagoons (e.g., Chattanooga, Tennessee). Acquisition of additional land from contiguous properties is often required, and encroachment of floodplains and floodways may be a constraint. Nearby neighbors may require consideration of odor and/or noise control. Flow equalization facilities are occasionally constructed at sites adjacent to remote influent pumping stations, where land is more readily available (e.g., Tulsa, Oklahoma, Northside and Southside WRRFs). The configuration of the flow equalization facilities should be designed considering the available footprint. Cast-in-place and prestressed concrete above-grade tanks with 7.5- to 9-m (25- to 30-ft) side water depths have been used to minimize space requirements. Above-grade steel tanks have been used for smaller facilities. Examples of wet weather flow equalization facilities are provided in Table 10.1.

3.0 SIZING AND TYPE OF CONSTRUCTION

3.1 Peak Flow and Volume Requirements

Frequencies, magnitudes, and durations of wet weather conditions and associated wastewater flows and volumes must be determined. Design conditions will vary across the country. Some regulatory agencies require full containment of a particular magnitude of storm event (e.g., 12.7-mm [0.5-in.], 19.05-mm [0.75-in.], or 25-mm [1-in.] rainfall or meeting a set rainfall recurrence interval and duration (e.g., 1-year, 2-year, or up to 25-year, 24-hour duration storm). Some regulatory agency guidelines call for consideration of flow equalization facilities when peak to average dry weather flow ratios are above a set amount.

In communities with little growth potential or relatively stationary or declining flows, a thorough statistical analysis of historical flow data will be required. Sewer system modeling will often be required to supplement historical data in growing communities and where collection system modifications are being made to reduce peak flows to the WRRF (real-time control, storage tunnels, remote storage facilities, and remote overflow treatment facilities) (see Chapters 2 and 9).

Oversizing of flow equalization facilities for the following purposes should be considered (see Chapters 16 and 8):

- Routing of sidestream flows during wet weather events,
- Solids storage during wet weather events, and
- A few hours of flow storage in the event of a power outage.

When considering the first two aforementioned items, care must be taken to evaluate the need for mixing or aeration and potential odor issues. Separate cells or basins for these uses may be more prudent.

3.2 Modular Design

Smaller, more frequent events may be handled in first-stage basins equipped with sludge removal equipment or automated solids flushing systems to make cleanup easier. Existing unused primary sedimentation basins are sometimes incorporated to this capacity. Multicell compartments, sometimes using cells of differing sizes, are sometimes provided in larger equalization basins to minimize cleanup requirements after small- to medium-size events, which are much more frequent than larger events.

TABLE 10.1 Example equalization facilities' summary descriptions.

Component	Farmington, Michigan[1]	Four Seasons Complex Lawrence, Kansas[a]	Northridge Montgomery County, Ohio[a]	Riverside Montgomery County, Ohio[a]	Northside (Mingo Creek) Tulsa, Oklahoma[a]
Total volume and peak flow	12.1 ML (3.2 mil. gal) 73.4 ML/d (19.4 mgd)	9.5 ML (2.5 mil. gal) 31.4 ML/d (8.3 mgd)	10.6 (ML) 2.8 mil. gal 76 ML/d (20 mgd)	2.6 ML (0.7 mil. gal) 38 ML/d (10 mgd)	137.0 ML (36.2 mil. gal) 193 ML/d (51 mgd)
Preliminary treatment	Two mechanically cleaned bar screens at 36.7 ML/d (9.7 mgd) ea.	Manual bar rack	None	None	Manually cleaned bar screens on top of overflow weir
Influent diversion/ pumps	Four nonclog centrifugal dry pits; 25.7 ML/d (6.8 mgd) ea. at 14 m (45 ft)	Three nonclog submersibles; two 17.4 ML/d (4.6 mgd) ea. at 22 m (72 ft); one 17.4 ML/d (4.6 mgd) at 9 m (30 ft)	Two adjustable side overflow weirs, 8 m (26 ft) long; five nonclog submersibles; 24.6 ML/d (6.5 mgd) ea. at 11 m (37 ft)	None	Side overflow weir; five nonclog submersibles; 48.4 ML/d (12.8 mgd) ea. at 13 m (42 ft)
Equalization basins	Cast-in-place concrete rectangular covered with six cells	Cast-in-place concrete open circular; 49 m (160 ft) dia. × 5.0 m (16.5 ft) sidewater depth	Prestressed concrete circular with concrete dome; 47 m (155 ft) dia. × 5.5 m (18 ft) sidewater depth	Prestressed concrete circular with concrete dome; 27 m (90 ft) dia. × 4.6 m (15 ft) sidewater depth	Concrete-lined lagoon with two cells
Type	Below-grade	Below-grade	Above-grade	Partially buried	Open lagoon

Mixing/ washdown systems	Spray water washdown system	Water cannons	Jet mixing and water cannons	Jet mixing and water cannons	Perimeter spray wash and hydrants with water cannons
Effluent control	Motor-operated eccentric plug valve	Motor-operated eccentric plug valve	Motor-operated eccentric plug valve	Two nonclog submersible; 4.9 ML/d (1.3 mgd) ea.	Motor-operated eccentric plug valve
Odor control system	Activated carbon; one 3.7 m (12 ft) dia. dual bed	Liquid-phase ferrous chloride addition	Activated carbon; two 3.7 m (12 ft) dia. dual beds	Activated carbon; one 3 m (10 ft) dia. dual bed	Liquid-phase ferrous chloride addition
Electrical/ instrumentation	500-kW standby generator	No standby power	No standby power	No standby power	No standby power

TABLE 10.1 Example equalization facilities summary descriptions (*continued*).

Component	Southside (West Bank) Tulsa, Oklahoma[a,b,c]	Bay View WRRF Toledo, Ohio[d]	Norman M. Cole, Jr., WRRF, Fairfax County, Virginia[e]	Snapfinger Creek WRRF, DeKalb County, Georgia[f]
Total volume and peak flow	129 ML (34 mil. gal) 129 ML/d (34 mgd)	95 ML (25 mil. gal) 700 ML/d (185 mgd)	61 ML (16 mil. gal)	76 ML (20 mil. gal)
Preliminary treatment	Two mechanically cleaned bar screens at 65 ML/d (17 mgd) each	13-mm (0.5-in.) clear spacing mechanical screens and vortex grit removal	Flow is diverted following primary treatment	—
Influent diversion/ pumps	Three nonclog centrifugal dry pit; 65 ML/d (17 mgd) each at 10.4 m (34 ft)	Remote 340-ML (90-mgd) pumping station and the facility influent PS	Four 149-kW (200-hp) submersible pumps with variable frequency drives	—
Equalization basins	Cast-in-place covered concrete square 3.8-ML (1.0-mil. gal) basin and 125-ML (33-mil. gal) concrete-lined lagoon with four cells	Cast-in-place rectangular concrete; two 15-ML (4-mil. gal) covered cells and a 65-ML (17-mil. gal) open cell; 7.6 m (25 ft) sidewater depth	Cast-in-place rectangular concrete with 4- to 15-ML (4-mil. gal) cells	61 m (200 ft) dia., 14.2 m (46.7 ft) (sidewater depth) with aluminum dome cover
Type	Buried basin and open lagoon	Partially buried	Partially buried	Partially buried
Mixing/ washdown systems	Floating aerators, perimeter spray wash, and water cannons	Siphon (vacuum) flushing chambers	Innovative precast concrete floor with semicircular furrows, sloped at 4%	—

Effluent control	Motor-operated eccentric plug valve	Motor-operated eccentric plug valve and 227-ML/d (60-mgd) low lift pumping station	—	—
Odor control system	Activated carbon for lift station and covered basin; seven 3.7-m (12-ft) dia. dual-bed units; liquid-phase sodium hypochlorite addition	Can be added to ventilation system on the covered portion in the future; has not been necessary	None	—
Electrical/ instrumentation	No standby power	Connected to facility standby power system	—	—

[a]Keller et al., 2000.
[b]Charron et al., 2012.
[c]Teusch et al., 2012.
[d]Lyon, 2005.
[e]Chapin et al., 2000.
[f]Walker and Heath, 2008.

3.3 Type of Construction

Construction includes new or converted conventional reinforced concrete treatment basins, which may be equipped with sludge collectors or wash-down systems; lined ponds or lagoons (reinforced concrete, asphalt, elastomeric compounds); prestressed concrete tanks; or above-grade steel tanks. The type of construction may be dictated by whether there will be mixing systems or odor control systems (requiring at least partial covers). Environmental conditions (e.g., extreme heat and UV exposure, freezing, freeze–thaw) greatly influence the selection of a lining system. If elastomeric lining materials are being considered, such as high-density polyethylene, EPDM, polyvinyl (vinyl chloride), or Hypalon, existing installations that have been in service in a similar climate for 20 or more years should be consulted regarding long-term serviceability and maintenance requirements.

4.0 PRETREATMENT REQUIREMENTS

4.1 Screening

The degree of screening required depends on the flow diversion mode and location and downstream process requirements where flow is returned. It is typically preferable to add screens to an existing headworks (by expansion, if necessary) to consolidate screenings handling and storage operations. Flow equalization facilities typically only require coarse screening to remove larger objects and rags that may interfere with mixing or aeration equipment. However, downstream biological processes, such as membrane activated sludge systems, require fine, omnidirectional screening (see Chapter 13, “Biological Treatment”). In some instances, equalized flow may be so dilute that it allows disinfection and direct discharge. In these cases, finer screening will be required to enhance disinfection effectiveness.

4.2 Grit Removal

Similar to screening, it is typically preferable to add units to an existing headworks (by expansion, if necessary) to consolidate grit handling and storage operations. If flow is diverted to flow equalization over a weir or from wet weather pumps from an overflow type wet well, then minimal grit removal may be required. A simple wide spot in the flow or a grit sump can be provided, with grit being periodically removed using a Vactor truck or similar mobile equipment. If grit removal is not provided in advance of flow equalization, then basin cleanup flows should be directed upstream of existing grit removal facilities.

5.0 OPERATION

5.1 Flow Diversion and Startup

Flow diversion can be passive (flow over a fixed weir) or active but should be automatic. Control methods are more fully described in Chapter 6. Measurement of equalization facility influent flow is not typically required but is beneficial for calculating peak influent flows, for recording excess flow that is treated and discharged, for pacing disinfection chemical feed for excess flow that is treated and discharged, or for evaluating flow equalization performance. Basin mixing systems or sludge collectors should be set to come on when minimum filling water levels are reached and to shut off when those minimum levels are reached again during basin dewatering operations.

5.2 Sustained Operation

Sustained operation refers to the situation in which at least the first stage of flow equalization is full and any mixing equipment, aeration equipment, sludge collectors, or odor control facilities are in operation. During sustained operation, periodic visual checks of the equalization facilities should be made to detect any special conditions requiring operator attention, including malfunctioning equipment, erroneous level readings, odor excursions, and foaming.

5.3 Operation as a Retention Treatment Basin

If a facility's National Pollutant Discharge Elimination System (NPDES) permit allows treatment and discharge from the equalization basin during wet weather (typically requiring disinfection), then both the design and operation are affected. The equalization basin should have a quiescent settling zone and overflow launders to minimize the surface overflow rate. Several two-sided weirs parallel to the direction of flow (i.e., "finger" weirs) are often used to increase the weir length and to minimize short-circuiting. Scum baffles should be provided to prevent floatables and scum from being discharged.

Retention treatment basins are most often configured for plug flow hydraulic conditions, especially if they are also used to provide chlorine contact time. Baffle walls are often added to provide a serpentine flow pattern in the chlorine contact zone, sized for a minimum of 10 to 20 minutes of detention time at the maximum hourly flowrate (or as otherwise required by state regulations and the NPDES permit). High-rate induction-type mixers are recommended at the entry to the chlorine contact zone to introduce and disperse the chlorine. Greater detail regarding disinfection of wet weather flows is provided in Chapter 15.

Retention treatment basins are often used for CSO treatment remote from WRRFs. For a more complete discussion of these types of facilities, consult the following references: *Prevention and Control of Sewer System Overflows* (WEF, 2011) and U.S. Environmental Protection Agency CSO Technology Fact Sheet: Retention Basins (1999).

5.4 Dewatering

Dewatering of equalization facilities should be initiated as soon as adequate treatment capacity is restored, particularly when more wet weather conditions are forecasted. Minimizing the holding time also mitigates odor formation. Flow equalization without mixing or aeration typically results in the equivalent of at least primary treatment so the stored flow can be returned to biological treatment. Equalized flow that is continuously mixed may need to be returned ahead of primary treatment for solids removal. Equalized flow that is aerated to maintain a minimum dissolved oxygen concentration of at least 1 mg/L may be sufficiently treated to be disinfected and directly discharged to the receiving waters if allowed by a WRRF's NPDES permit. Refer to the discussion in Section 5.3 on retention treatment basins.

5.5 Cleanup and Preparedness for the Next Event

Most flow equalization facilities will require cleanup after each use, even if provided with continuous mixing or aeration. Mixing and aeration equipment typically needs to be shut off at a fixed low-water cutoff elevation, allowing some residual solids to settle. After the equalization basins have been dewatered, flushing systems should be activated to wash down any accumulation of solids. Flushing or washdown flows should be directed to preliminary or primary treatment. If grit removal is not provided ahead of flow equalization, it may be necessary to return cleanup flow ahead of the WRRF's grit removal facilities. Those portions of flow equalization facilities that are completely covered and provided with odor control facilities may be cleaned at slightly lower frequencies to avoid confined space entry requirements.

6.0 SIGNIFICANT COMPONENTS

6.1 Mixing and Aeration Systems

Some state regulatory guidance documents for wastewater treatment recommend mixing or aeration systems for flow equalization. There are a few situations where equalization basin mixing or aeration is beneficial, such as the following:

- Warm weather first-flush-type events,
- Discharges from long force mains that have become septic,
- Long flow equalization detention times exceeding 2 or 3 days,
- Where high-strength wastes are received, and
- Where flow equalization contents may be discharged without going through secondary treatment.

Mixing and aeration systems most commonly used in flow equalization applications include the following: floating aerators, submersible mixers, jet mixing systems, and diffused aeration.

Floating aerators are probably the most commonly used mixing and aeration systems for flow equalization, particularly in large open lined lagoons. The primary advantage of floating aerators is that they provide minimal disruption for cleaning solids from the bottom of the equalization basin. Design guidelines for mixing equipment call for 0.0005 to 0.001 kW/m^3 (0.02 to 0.04 $hp/1000\ ft^3$) of basin volume (GLUMRB, 2004; U.S. EPA, 1974). The standard oxygen transfer efficiency of floating aerators is typically in the range of 1.2 to 1.8 kg $O_2/kW{\cdot}h$ (2 to 3 lb $O_2/hp{\cdot}h$). The actual transfer efficiency in wastewater may be one-half or less than that in clean water. Equipment manufacturers should be consulted for specific details and performance expectations. Flow equalization basins equipped with floating aerators typically require additional freeboard than nonmixed basins because of the waves they create and surface spray.

Submersible mixers work well in circular basins, basins emulating a racetrack configuration, or other plug flow (unidirectional flow) basins. Maintaining a minimum velocity of approximately 0.3 m/s (1.0 ft/s) in the direction of flow is the primary objective. In circular basins, an inlet located tangentially at or near the bottom of the basin is effective at starting unidirectional flow that is reinforced by submersible mixers that propel the flow in the same direction. This arrangement has worked well at the Lincoln Park CSO storage facility in Decatur, Illinois, resulting in a relatively self-cleaning basin.

Jet mixing systems use submerged nozzles to disperse recirculated flow and air to impart mixing of the basin contents. Combination headers are used with the air supply piping installed inside of the flow recirculation piping. A smaller airflow nozzle discharges through a larger flow recirculation nozzle at several locations along the header, which is mounted transverse to the desired direction of flow in the basin. Nonclog wastewater pumps are used to recirculate the flow and air blowers typically provide the mixing air supply. Naturally aspirating designs are also available for installations that do not require precise control of the air supply (e.g., for mixing, but not for aeration). Screening is required with jet mixing systems to the degree necessary to keep from clogging the flow recirculation nozzles. Equipment manufacturers should be consulted for design details, layout requirements, and performance expectations.

Diffused aeration systems are seldom used in offline flow equalization facilities because the bottom-mounted air diffusers interfere with basin cleanup. They are more commonly used in inline equalization basins that are used daily to dampen peak flows and loadings to downstream processes. Stainless steel coarse bubble diffusers, or similarly robust diffusers that can withstand weathering in a wet–dry environment, should be used.

6.2 Dewatering Pumps/Control Valves

The type of dewatering pumps to use is dictated by the pretreatment that is provided. If coarse or no screening is provided, it may be advisable to use chopper pumps. Variable suction head conditions as basins are drawn down present some challenges in pump selection and in controlling the pump discharge rate. Partial gravity dewatering may alleviate this condition somewhat. Carefully evaluate the entire range of operation of the dewatering pumps so that a proper motor size is provided. A dewatering sump at least 1 m deep will improve pump suction conditions and allow for more complete dewatering of basins.

Motor- or pneumatic-operated eccentric plug valves in conjunction with downstream flow meters (magnetic on ultrasonic type) are most often used for controlling gravity return flows to the WRRF. Several pipe diameters should be provided upstream of the flow meter for more precise flow control. The control valve, flow meter, and a downstream shutoff valve are typically installed in a below-grade vault to facilitate periodic maintenance. Confined space entry requirements must be observed when entering the vault.

6.3 Basin Cleaning Systems

6.3.1 Sludge Collectors

In multiple basin flow equalization facilities, conventional primary sedimentation basins (either existing or new) are often used as the first stage. In these types of basins, traditional sludge collectors may be used to facilitate basin cleanup. If influent enters the basin over a weir, baffle wall, or through a suspended pipe, then provisions should be made to prevent influent flows from cascading onto the clarifier equipment during basin filling.

6.3.2 Tipping Buckets

Tipping buckets are used in rectangular basins in conjunction with gently sloping floors (2% minimum slope), training walls to contain the traveling wave of flushing water, and a receiving sump. The receiving sump must allow for free discharge, and should have a volume at least 10% greater than the volume of the tipping bucket. Tipping buckets are mounted on wall brackets at the shallower end of the basin and, when filled with flushing water, automatically rotate to empty their contents down the wall and onto the basin floor. A formed, curved transition at the base of the wall forms the flow into a traveling wave of flushing water that travels the length of the basin floor. Tipping buckets are effective at lengths of up to 40 m (130 ft) or more, depending on the height of the tipping bucket. Treated WRRF effluent or potable water are most often used to fill the tipping buckets. Clear flushing lanes, free of equipment or structural obstacles, should be provided with tipping buckets or any of the other flushing systems described herein. Refer to Figure 10.2 for an example of basin flushing systems.

6.3.3 Hydraulic Flush Gates

Hydraulic flush gates are similar in concept to tipping buckets, but use a reservoir of water held back by a hydraulically actuated flap gate to provide the flushing water. The reservoir of water may be of any shape, allowing this type of flushing system to be used in circular tanks (see Figure 10.2). Hydraulic flushing gates have a larger volume of flushing water and release the flushing water in a "dam break" type of scenario; thus, have a slightly longer range of flushing length than tipping buckets, especially in shallower basins.

6.3.4 Siphon Chambers

Siphon chambers are similar in concept to tipping buckets, but have a water reservoir created by a P-trap and vacuum chamber at the shallow end of the basin. The reservoir chamber is filled as the basin fills and then a vacuum is applied to the chamber before dewatering. The P-trap aids in maintaining the vacuum while the basin is being dewatered. When dewatering is complete, the vacuum is released and the reservoir contents flush the basin floor (see Figure 10.2). Siphon flushing chambers have a slightly longer flushing range than either tipping buckets or flushing gates resulting from their greater volume of flushing water. Multiple siphon chambers are used

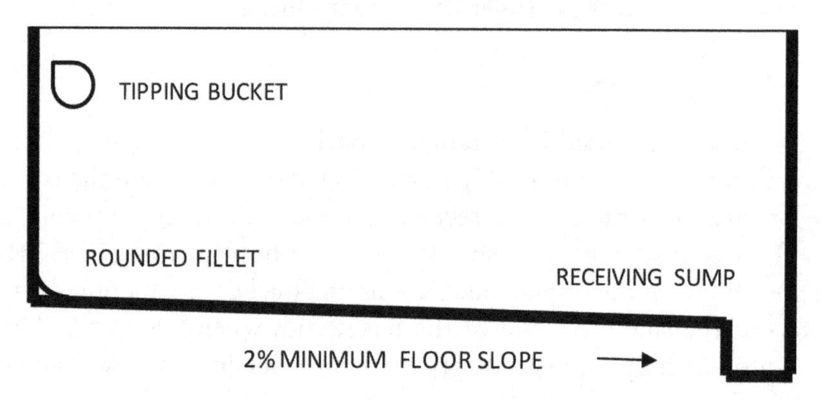

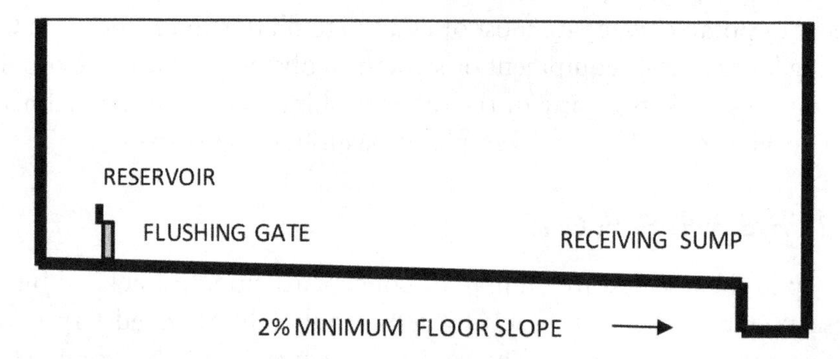

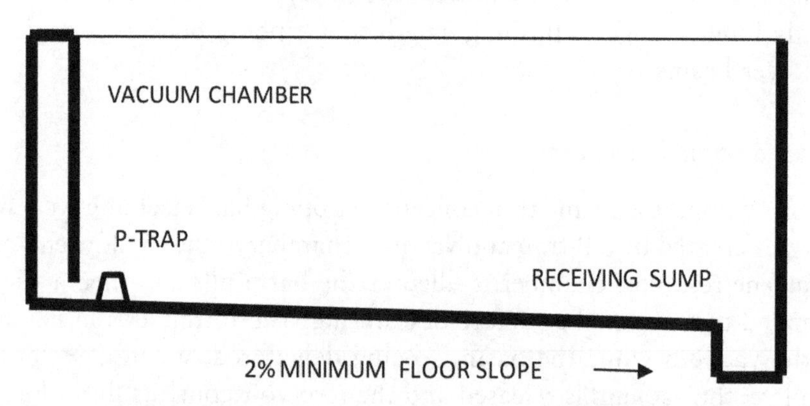

FIGURE 10.2 Basin flushing systems.

in the 95-ML (25-mil. gal) equalization basin at the Toledo, Ohio, Bay View WRRF. Occasional difficulties associated with maintaining the vacuum in the relatively large chambers surfaced after a few years of operation. Extra care is required to ensure complete sealing of the chambers during design and construction and as a preventive maintenance measure. A unique application for a siphon chamber is to construct it as a cylinder in the center of a circular basin with a perimeter sump to capture the flushing water. The flushing water empties radially to clean the basin floor.

6.3.5 Water Cannons/Yard Hydrants

Water cannons and yard hydrants with fire hoses are commonly used for circular concrete basins or concrete-lined basins. When fixed water cannons are used, the floor should be designed to slope away from the water cannon locations to the maximum practicable extent. In colder climates, self-draining pipes and/or frost-proof hydrants must be used where piping is exposed to the elements for freeze protection.

6.3.6 High-Pressure Spray Water Systems

High-pressure (typically around 345 to 414 kPa or 50 to 60 psig) spray water systems are sometimes used for basin cleanup, especially in hard-to-access areas. Yard hydrants are often furnished for supplementary cleaning in conjunction with spray water systems. In most instances, spray water systems are not as effective as the more turbulent flow washdown systems described previously.

6.3.7 Manual Cleaning

For large, open lined lagoons, it is often the practice to provide access ramps for Bobcats or similar small equipment to enter the basin and scrape solids toward collection troughs or sumps for basin cleanup. Final cleanup is typically accomplished using fire hoses connected to yard hydrants. These measures may be required because of the physical limitations of alternative cleaning systems that may not have a sufficient cleaning range.

6.4 Odor Control

Complete odor control systems are seldom required. Many installed systems are not being used or are only occasionally being used. Odor control for a first-flush-type of tank, particularly if receiving flow from a long force main, may be necessary. If wastewater is to be stored for a few days, then odor control facilities should be considered. A common practice is to make provisions for adding basin covers and odor control in the future in case odor excursions occur.

7.0 REFERENCES

Chapin, J.; Hogge, A.; Gupta, K. (2000) New Pump Station and 16 MG Equalization Tank for Peak Flow Shaving at the Norman M. Cole, Jr. Pollution Control Plant, Fairfax County, Virginia. *Proceedings of the 73rd Annual Water Environment Federation Technical Exposition and Conference* [CD–ROM]; Anaheim, California, Oct 14–18; Water Environment Federation: Alexandria, Virginia; Session 51–60, pp 385–403.

Charron, A.; Belanger, F.; Vaughn, M. (2012) Efficient Control Improvements Increasing the Use of Flow Equalization Basins. *Proceedings of the 85th Annual Water Environment Federation Technical Exhibition and Conference* [CD–ROM]; New Orleans, Louisiana, Sept 29–Oct 3; Water Environment Federation: Alexandria, Virginia; Session 108.

Great Lakes Upper Mississippi River Board of State and Provincial Public Health and Environmental Managers (2004) *Recommended Standards for Wastewater Facilities;* Health Research, Inc.: Albany, New York.

Keller, J.; Zilla, R.; Lyon, T.; Corwin, B. (2000) How Much Is that SSO Equalization Facility Going to Cost? *Proceedings of the 73rd Annual Water Environment Federation Technical Exposition and Conference* [CD–ROM]; Anaheim, California, Oct 14–18; Water Environment Federation: Alexandria, Virginia.

Lyon, T. (2005) Comprehensive Evaluation Solves Wet Weather Treatment Challenges. *Proceedings of the 78th Annual Water Environment Federation Technical Exhibition and Conference* [CD-ROM]; Washington D.C., Oct 29–Nov 2; Water Environment Federation: Alexandria, Virginia.

Teusch, J.; David, R.; Shelton, R.; Barton, T. (2012) Tulsa's Storage Solution to SSO Reduction. *Proceedings of the 85th Annual Water Environment Federation Technical Exhibition and Conference* [CD-ROM]; New Orleans, Louisiana, Sept 29–Oct 3; Water Environment Federation: Alexandria, Virginia; Session 18.

U.S. Environmental Protection Agency (1999) *Combined Sewer Overflow Technology Fact Sheet: Retention Basins;* EPA-832/F-99-042; U.S. Environmental Protection Agency: Washington, D.C.

U.S. Environmental Protection Agency (1974) *Flow Equalization: EPA Technology Transfer Seminar Publication;* U.S. Environmental Protection Agency: Washington, D.C.

Walker, D.; Heath, G. (2008) Design Considerations for Off-Line Storage Tanks. *Proceedings of the 81st Annual Water Environment Federation Technical Exhibition and Conference* [CD–ROM]; Chicago, Illinois, Oct 18–22; Water Environment Federation: Alexandria, Virginia.

Water Environment Federation (2011) *Prevention and Control of Sewer System Overflows,* 3rd ed.; Manual of Practice No. FD-17; Water Environment Federation: Alexandria, Virginia.

8.0 SUGGESTED READINGS

Geisser, D.; Capozza, N.; Perriello, J. (2011) Cleaning of CSO Facilities Using Various Flushing Devices—A Case Study for the Midland Avenue Regional Treatment Facility, Onondaga County, New York. *Proceedings of the 84th Annual Water Environment Federation Technical Exhibition and Conference* [CD–ROM]; Los Angeles, California, Oct 15–19; Water Environment Federation: Alexandria, Virginia.

Leffler, M.; Harrington, J. (2001) SSO Elimination Through Expanded Primary Treatment Capacity and Blended Effluent. *Proceedings of the 74th Annual Water Environment Federation Technical Exposition and Conference* [CD–ROM]; Atlanta, Georgia, Oct 13–17; Water Environment Federation: Alexandria, Virginia; Session 61–70, pp 49–61.

Longaria, R.; Cleveland, P.; Brashear, K. (2006) Achieving Reduced-Cost Wet Weather Flow Treatment Through Operational Changes. *Proceedings of the 79th Annual Water Environment Federation Technical Exhibition and Conference* [CD–ROM]; Dallas, Texas, Oct 21–25; Water Environment Federation: Alexandria, Virginia.

U.S. Environmental Protection Agency (1979) *Evaluation of Flow Equalization in Municipal Wastewater Treatment;* EPA-600/2-79-096; U.S. Environmental Protection Agency: Washington, D.C.

11

Preliminary Treatment

William Eleazer, P.E., and Allen P. Sehloff, P.E.

1.0 SCOPE

The primary purpose of preliminary treatment is the removal of debris and grit from the influent wastewater. Debris can clog pumps, pipes, and other downstream equipment while grit can cause premature wear on pumps, collector flights, centrifuges, and other process equipment and can collect in tanks and channels, thus reducing capacity. Wet weather events frequently result in increased screenings and grit loads on a water resource recovery facility (WRRF). During typical flow periods, debris and grit can settle and accumulate in collection system piping, pumping stations, and within the facility itself. When high flows from wet weather events occur, these materials are fluidized and are transported to the WRRF. With combined collection systems, wet weather can bring to the WRRF an additional large influx of debris and grit that is not experienced during typical flow periods. This chapter focuses on effects, design, and operational considerations associated with preliminary treatment systems during wet weather events.

1.1 Screenings Removal

It is preferable to remove large debris ahead of grit removal to prevent clogging of grit removal equipment. There are more than a dozen types of screens that can be used at a WRRF. Depending on the size of the WRRF, the nature of the incoming debris, and the downstream processes at the facility, any number of theses screens may be used or may be candidate screens for future use at the WRRF. For new installations, a preliminary review of available types of screens vs intended service is initially performed to narrow the candidate screen types to a reasonable subset (two to three) for further study. A more detailed analysis follows that includes a study of costs, efficiencies, and general preferences. Although the screen aperture size selection for a given WRRF will vary, it is increasingly common for facilities installing new screening systems to select aperture sizes of 10 mm or smaller, with 6 mm becoming increasingly common and 3 mm being less common, but gaining popularity. If the downstream aeration process includes moving bed biofilm reactors (MBBR)/integrated fixed-film activated sludge (IFAS) systems, 3-mm screening is increasingly becoming more common. If the downstream process includes membrane biological reactors, screening systems with aperture sizes of 3 mm and less are required and typically require at least two stages of screening.

Screenings are typically discharged to a receiving dumpster for temporary storage and eventual transport to final disposal, which is typically a landfill or incinerator. The method of conveying screenings from the screen to the dumpster vary widely and are dependent on the size of the WRRF, the nature of the collection system and the debris it transports, the screen and screen aperture chosen, the preliminary treatment area layout, and the desired end quality of the screenings. The selection of the conveying system is performed during an options analysis conducted during preliminary design. Since the mid-1990s, the use of washer/compactors integral to the conveying system has become nearly standard practice, especially with noncombined sewer collection systems and when the screening aperture sizes are 10 mm and smaller. This largely results from the increased popularity of smaller aperture sized screening through this same time frame, that these screenings include an increased amount of organics, and the requirement from landfills that screenings not include free water. Additional attention has also been placed on operations costs in which fourfold decreases in hauling costs for screenings have been realized with the installation of washer/compactors (Casey et al., 2009).

1.2 Grit Removal

The most common types of grit removal currently in use are aerated grit removal and vortex grit removal with a tangential inlet, either mechanically enhanced or with multiple trays. Aerated grit tanks are custom designed to meet the needs of a WRRF, whereas vortex units are typically rated by the equipment manufacturer for a peak hydraulic loading rate and a minimum capture rate for grit at a particular size. Actual performance of any grit removal unit will vary based on site-specific grit characteristics.

Grit is typically pumped from the removal units using severe-duty recessed impeller pumps. The grit slurry is then concentrated using some type of cyclonic separator and dewatered using a grit classifier. There are several manufacturers of this type of equipment and the designer must select the system that meets the needs of a given facility. Some of this equipment focuses on producing a clean, dry grit with an organics content of less than 5% by weight, whereas others focus on the capture of grit down to a particle size of 75 µ. The definitions of *coarse*, *medium*, and *fine grit* may vary, but, for the purposes of this publication, *fine grit* is defined as smaller than 105 µ. In addition to particle size, the specific gravity of the grit, the coating of grit particles with organic material such as grease, and many other factors will affect grit capture. As a general rule, the capture of fine grit is sacrificed when a clean, dry grit with a low organics content is selected as the basis of design, while the capture of fine grit particles (≤105 µ) typically results in a wetter grit with a higher organics content.

1.3 Flow Equalization

Chapter 10 provided details regarding flow equalization. Flow equalization is frequently included with preliminary treatment and may be located upstream of screening facilities or downstream of grit removal. The location of flow equalization will affect the sizing of preliminary treatment facilities.

2.0 EFFECTS OF WET WEATHER FLOWS ON PRELIMINARY TREATMENT

There are two significant effects on the preliminary treatment system during wet weather. The first is increased flow, which can dramatically increase during wet weather. The second is increased debris and grit load during the first flush phenomenon, which typically accompanies the first phase of a wet weather event. During normal flow periods, grit and debris can settle and accumulate in collection system piping and pumping stations. This phenomenon is also prevalent within the WRRF itself, where large, diameter pipes, transfer channels, wet wells, and manholes can accumulate solids at their inverts and floatable debris at their crowns or on the water surface. As flows increase and reach their peak, these deposited solids and accumulated debris are fluidized or dislodged and are carried downstream, frequently as a slug load. The first flush phenomenon is significantly more dramatic in combined sewer systems and these systems will experience an even more pronounced influx of debris and grit during the first phase of a wet weather event. The deposition of solids in channels and tanks within the WRRF can also lead to problems with flow-distribution characteristics within the facility and the capacity of the facility. Sufficient deposition can occur such that the desired flow distribution goals cannot be achieved. In addition, sufficient accumulation can occur in tanks such that the treatment capacity of the tank may be sacrificed. It should be noted that the first flush phenomenon can occur daily as well, although to a much less extreme, and be related rather strictly to diurnal peaking (Wodrich, 2005).

The typical pattern for incoming debris and grit to the preliminary treatment system during a wet weather event is such that, during the first portion of the event, there is a significant increase in the influent concentration (kilograms per liter) and mass (kilograms per minute) of debris and grit. This results from the first flush phenomenon. The magnitude and length of this peak is dependent on the characteristics of the collection system and the wet weather event itself. The first flush can lead to quick overloading of the preliminary treatment systems. Following the first flush, while the high flow continues, there is typically a significant drop in the influent concentration of debris and grit; however, the mass of influent debris and grit per minute will tend to track with normal flow tendencies. Each system is unique and careful study is required on a facility-by-facility basis to determine appropriate influent design values.

2.1 Effect on Screenings Systems

2.1.1 Screens

Screen performance is defined by the screen retention value (SRV), which is the percentage of the total influent solids retained by the screen. The SRV will have a tendency to decline with increasing approach velocities. As approach velocities increase to more than 1 m/sec during high flow, several factors lead to deteriorated performance. The first and most pronounced is the tendency of debris to be pushed through the screen's apertures with the high force caused by the increased velocity that occurs during the high flows accompanying wet weather events. The second factor is the physical disruption of the static barriers (i.e., plates, brushes, flaps) at the interface between the screen and the channel walls and floors and between the screen frame and the screening elements and the same push-through effect occurring at these locations. Even when small aperture-sized screens (3 mm) are installed, much larger debris can pass through these screen resulting from the push-through effect at the annular spaces of screens (Koch and Hielema, 2010).

Most screens have a certain fraction of carryover of screenings. This carryover directly affects a screen's SRV and it is composed of debris that was initially captured on the screen face, but was carried over to the discharge channel. Because of the increased mass of influent screenings during the first flush, there will be a tendency for any given screen to experience an increase in the mass of carryover during the first phase of a wet weather event.

The increased grit load that can accompany a first flush can severely affect the screening system. Because the aperture velocity is greater than the channel velocity, a sudden headloss occurs at the screen face. At this location, there is a tendency for grit to accumulate and mound. The first flush can be accompanied with a tremendous influx of grit and this grit can rapidly accumulate at the screen face. This can have catastrophic implications on the screen causing the screen cleaning rakes or moveable plates to bind and seize, and rendering the cleaning of the screen elements inoperable. This subsequently can lead to the near immediate blinding of the screen, a rapid rise in the upstream channel water surface elevation, and the need to bypass flow around the screen. Significant maintenance activities will be required following the event to shut down the channel, remove the accumulated debris and grit, and repair any damage that may have occurred. Figures 11.1 and 11.2 show accumulated debris and grit upstream of a screen following a wet weather event.

In many areas, a wet weather season occurs simultaneous with the leaf-drop season. With combined collection systems, the coupling of these two occurrences can have negative effects on preliminary treatment systems at a WRRF. The potential of leaf blinding of screening systems caused by autumn

FIGURE 11.1 Digging out accumulated grit following screen failure during a first flush event (courtesy of Brown and Caldwell).

FIGURE 11.2 Accumulated debris following a first flush event (courtesy of Brown and Caldwell).

rains coupled with the leaf-drop season for systems that receive flows from combined sewers require special attention by the designer.

2.1.2 Channel Velocity and Hydraulics

Two critical velocity parameters exist with screening systems. The first is the approach velocity and the second is the aperture velocity. The approach velocity is the velocity of the water column within the approach channel to the screen. The aperture velocity is the velocity through the apertures in the screening element.

Under the range of typical operating conditions, an attempt should be made to design the screening channels to maintain an approach velocity greater than the deposition velocity. This will minimize sedimentation in the channel. There is no minimum approach velocity that leads to reduced screen performance and, in general, the lower the approach velocity, the better the screen will perform with the important caveat of maintaining sufficient velocity to prevent significant sedimentation in the channel. The deposition velocity of fine grit is significantly lower than that of larger grit (pebbles, small stones) and, as such, the target velocity in the channel is used to balance the deposition of both of these types of grit. A reasonable target approach velocity for a screenings channel is 0.5 m/sec. This represents the velocity at which larger sediment will begin to fall out of suspension in the channel and accumulate. In instances where the approach velocity drops well below 0.5 m/sec and for extended periods of time (several hours) (e.g., during diurnal low flow periods), the channels should be designed to periodically (at least daily) achieve an approach velocity greater than the resuspension velocity. For normal wastewater, this value is typically cited to be 1.0 m/sec for screening channels; however, some design professionals prefer to design this at 1.5 m/sec, with the belief that the heavier grit that has settled in the channels requires a much higher velocity to resuspend. These parameters are especially critical during the first flush. During the first flush, if the approach velocity is not above the sedimentation velocity, significant problems associated with grit sedimentation can result. Grit sedimentation can cause imbalances in the flow to the screens and, if there is enough sedimentation at the screen face, the screens can experience catastrophic problems associated with failures in their cleaning devices as described earlier in this section. As noted earlier in this chapter, an approach velocity of greater than 1.0 m/sec will begin to undermine the screen's effectiveness. Where a choice is required, especially when it concerns the high flows during wet weather, it is typically better for the designer to err on the side of higher velocities to maintain the approach channels free of grit deposition and prevent catastrophic problems with screen functionality.

The aperture velocity is the velocity of fluid through a clean aperture. Because the aperture is noted as being clean, there is no aperture velocity that causes reduced or enhanced screen performance. Push-through velocity is correlated with approach velocity and not aperture velocity. The aperture velocity is only used to calculate the headloss across the screen. Maximum recommended manufacturer aperture velocities are sometimes provided; however, those are typically provided to minimize excessive headloss across the screen.

Headloss across a screen will dramatically rise with increased flow because headloss is a function of the square of the aperture velocity. This relationship causes the water surface level upstream of a screen to rise rapidly on the increasing flow that may occur during a wet weather event. The two concerns with headloss are influent channel overflow and screen damage. As flows increase, the headloss and the related water surface level upstream of the screen increase dramatically because of the exponential nature of their relationship. This can cause the channel to overflow. Bypass systems are important, with passive and rapid engagement bypasses being optimal. The headloss is further related to the blinding percentage used in the headloss calculation. The blinding percentage is the assumed percentage of aperture area that is not available to pass fluid because it is not in a clean state. At any given time, a portion of a screen will be blinded; otherwise, it would not be capturing any debris. A 40% blinding percentage for most screen types is a typical design parameter for high-flow headloss calculations. If significant first flush debris is anticipated, particularly in the case of leaf loading, a 50% blinding percentage is recommended. The structural integrity of a machine is related to the amount of differential headloss that the machine can withstand without incurring damage. Requiring that the screen be capable of structurally withstanding, without damage, a differential headloss equivalent to the entire channel depth is recommended.

2.1.3 Ancillary Systems

There are four primary ancillary systems associated with screens: washing/compacting, conveying, storage, and supplemental supply water. Wet weather flows can be accompanied with larger and heavier debris and a much greater mass of debris. This is especially true during the first flush phase. Larger debris may have been accumulating within the collection system during low and normal flow periods or may have entered the collection system from the street level in a combined system. This larger debris and the overwhelming mass of debris that can accompany the first flush can jam washer/compactors and conveying machinery such as sluices and screw conveyors. Belt conveyors are less susceptible to jamming; however, they are not suitable for many types of screens. Similarly, storage facilities can become overwhelmed and, as a result, mounding and overflowing can occur. During wet weather events, the cleaning of the screens and the operation of the washer/compactors will be at a maximum as will their supplemental water requirements. These supplemental water requirements can outpace available supply, causing a drop in pressure in the system and a correlating reduction in the cleaning capabilities of screen types that use supplemental water to aid in the cleaning of their screen elements.

2.2 Effect on Grit Removal Systems

2.2.1 Influent Grit Quantities

Heavy grit loads may be anticipated during wet weather events as grit in the collection system and in channels upstream of preliminary treatment is fluidized and transported to the WRRF.

2.2.2 Channel Velocity and Hydraulics

Channel velocity and hydraulics are affected by the grit removal process selected for a given WRRF. Configuring influent channels to provide relatively uniform grit loadings to the grit facilities in service is always a challenge. Computational fluid dynamic modeling may be useful in identifying areas of low velocity in channels where grit deposition may occur. For large facilities, construction of a physical model may also be of value.

2.2.3 Grit Removal Units

Design of Municipal Wastewater Treatment Plants (WEF et al., 2010) provides information on incorporation of the various grit removal technologies in a WRRF process train. When selecting the appropriate grit removal technology for a WRRF with high wet weather flows, headloss under the range of anticipated flows must be considered. This is particularly true of existing WRRFs in which a new grit removal unit is being inserted within an existing hydraulic grade line.

2.2.3.1 Aerated Grit Removal

Of the grit removal alternatives commonly in use, hydraulic effects are probably the lowest for aerated grit removal. The greatest challenge with aerated grit removal is uniform distribution of both flow and grit among the units in service. A low headloss distribution channel with an entrance loss at each grit removal unit of at least 10 times the loss along the length of the channel at peak flow will provide reasonable flow distribution among the tanks in service; however, uniform hydraulic distribution will not guarantee uniform grit distribution among the facilities in service. Grit deposition in the distribution channel can also be a problem. Aeration or some other means of mixing should be provided to keep grit in suspension. If aeration is provided for distribution channel mixing, the inlets to the individual grit tanks should be flush with the bottom of the distribution channel and the channel aeration should be configured to set up a spiral roll pattern to sweep the grit into the tanks and minimize grit deposition in the distribution channel.

2.2.3.2 Vortex Grit—Mechanically Assisted

Mechanically assisted vortex grit removal units have a range of approach velocities, typically 0.5 to 1.1 m/sec (1.6 to 3.5 ft/sec). This velocity range is set by the manufacturer to minimize grit deposition in the approach channel and to set up a vortex in the grit removal unit. At velocities above the manufacturer's specified range, grit capture may decline.

2.2.3.3 Vortex Grit—Multiple Trays

Vortex units with tangential inlets and multiple trays (e.g., Hydro International Head Cell) also have a maximum capacity established by the manufacturer. At flows above the maximum, the headloss through the unit could be significant, potentially affecting upstream facilities; however, these units typically occupy less space than aerated grit and, as such, they may be viable if headloss is not a concern. The headloss through these units should be considered when analyzing the hydraulic profile.

2.2.4 Ancillary Systems

2.2.4.1 Grit Pumps and Piping

Grit pumps can be overwhelmed by grit during a wet weather event. If the pumps cannot keep up with the grit entering the hopper, the hopper will pack with grit and the system will fail. Pump suction pipe length should be kept as short as practical with minimum bends to minimize pump inlet piping losses, the possibility of pipe plugging, and the effort required to clear a plugged inlet pipe. High-pressure facility reuse water feed with a solenoid valve control to the grit pump suction lines timed to open on or just before pump startup will help fluidize compacted grit if pumping is intermittent.

2.2.4.2 Grit Separators

Hydro-cyclones and similar grit concentrators can clog with debris. This can happen under any flow condition, but is more likely during wet weather when large quantities of debris in the collection system are fluidized and transported to the WRRF and deteriorated performance of the screening process can occur. Grit causes wear to plastic liners in cyclones and the liners will require periodic (often yearly) replacement for optimal performance. Some manufacturers use hardened material such as Ni-Hard for construction of the hydro-cyclones.

2.2.4.3 Grit Classifiers

Multiple grit separators may discharge to a single classifier. Classifiers need to be sized for the anticipated peak grit and hydraulic loads. If a grit classifier is not sized for the peak grit and hydraulic loads and has not been optimized for performance, it may be overwhelmed and fail during a wet weather event.

2.2.4.4 Decant Tanks

Decant tanks with overflows for typical operation and wedge-wire bottoms for draining before disposal are used at some WRRFs. Some smaller facilities eliminate separators and classifiers entirely, pumping the grit slurry directly to a decanter. More efficient capture of grit may be possible using decant tanks than can be accomplished using separators and classifiers, and far less infrastructure is required. This needs to be considered during facility design.

3.0 DESIGN AND MODIFICATION CONSIDERATIONS

3.1 Screens and Screen Channels

Because of the large mass of debris associated with the first flush of a wet weather event and the overall stresses on the screening system resulting from the increased flow, it is particularly important for a screening process to maintain its overall functionality throughout the wet weather period. Preferably, debris should be removed from the incoming wastewater before the grit removal process because the debris in raw wastewater can clog the equipment within the grit removal system, rendering it nonfunctioning. There are, however, many existing facilities that were constructed with the grit removal process in front of the screening process. These facilities are particularly difficult to modify in a way that places the screening process in front of the grit removal process resulting from the extreme expense related to modifying the influent piping system, the overall flow pattern, and the hydraulic grade of the influent wastewater and through the preliminary treatment processes. Some facilities that were constructed with the grit removal process before the screening process are modifying their facilities by constructing a new first screening system before the grit removal process and retrofitting the existing screening system to be a second screening system. The new first screening system is constructed with 12- to 25-mm screens and the second screen is modified with 3- to 10-mm screens. This screening–grit removal–screening modification scenario enables the facility to minimally disturb the hydraulic profile (the first screens have minimal hydraulic-grade effect), reduce clogging issues of the grit removal process (typically driven by debris larger than 12 mm), and provide the additional screening that greatly enhances the overall performance of all downstream processes.

The hydraulic-grade implications of installing a screen with a smaller aperture size can be more complex than initially anticipated. Headlosses through the smaller-aperture-sized screens can result in headlosses 3 to 4 times or more than that of the existing screens, often surprising the design professional. Careful hydraulic analysis is required. Recent research (Botero et al., 2011) has indicated that the most common design formulas for estimating headloss through screens are inaccurate and typically overestimate the amount of headloss that will be experienced. This research found that computational fluid dynamic modeling more accurately predicted headloss across a screen. Careful review by the design professional and certification by the screen manufacturer are recommended. A common modification strategy is to add channels. Some facilities, however, are constrained and adding channels is not a readily viable option. Installing curbing around existing open channels leading to the screen is a method by which the hydraulic grade upstream of the screen can be raised, screens with a smaller aperture size can be installed, and the overall footprint of the screenings area is not significantly modified.

All screens that receive a significant wet weather flow and its accompanying first flush and all screens that receive flow from combined sewer collection systems should be installed with the fastest possible cleaning rate. Using rakes that have a high individual capacity is also recommended. Occasionally, existing screens can be modified to use a faster cleaning rate or multirake screens can be retrofitted with additional cleaning rakes. Considering both the speed and overall mass capacity, the cleaning rate of a screen is critical to preventing a screen from exceeding its assumed design blinding percentage and in maintaining the upstream water surface level below the top of the channel.

Installing screens at a lesser angle to horizontal provides more screen contact surface area with the water column and subsequently reduces the headloss across the screen for the equivalent cross-sectional channel dimension. This also reduces the overall energy profile of the screening system because the headloss across the screen is less. In many European installations, screens are installed at 45 deg, recognizing the energy benefit of the reduced angle. This, however, leads to additional initial investment because the overall length of the machine and channel and the footprint of the screening area are larger. Accordingly, in the United States, most screens are installed at 60 deg or more from horizontal.

Agitation air can be added to the screening channels to aid in resuspension of grit. This system can be run constantly or intermittently during low-flow periods, during first flushes of wet weather events, or as needed to aid in maintaining grit loads in suspension in the screening channels. Channels can also be shaped with a V-configuration to maintain velocities as flows are reduced. The V-configuration is difficult at the interface with the screen because the channel must return to full rectangular shape; however, agitation air can be added in this location should there be concern with grit accumulation at the face of the screen. Hydraulic ramps at the bottom of the channel can also be used at the face of the screen. These ramps attempt to lift the grit through the screening elements in place of accumulating at the bottom of the screen where the solid bottom piece is located and where there is no screen element.

Exceeding recommended approach velocities is common with screen systems that receive significant wet weather flow. It is difficult to balance the desired low-flow approach velocities and the recommended maximum approach velocity during high flow without constructing a significant number of small channels, which is typically not economically feasible. Accepting periodic low-flow velocities, using antisedimentation design tools (agitation air, V-shaped channels, and approach ramps), using operational strategies such as opening and closing channels and performing maintenance channel cleaning, and recognizing that recommended high-flow velocities may be surpassed and screening capture may be sacrificed at high flows constitute the balancing efforts that the engineer must perform during the design of a screening system.

Because screen channels are typically designed with their influent and effluent gates being the only means of isolation, the channels should be constructed with bulkhead cutouts ahead of and behind each of the isolation gates to facilitate isolation and repair of these gates.

The use of rock boxes, or rock traps, is discussed later in this chapter, although they have been implemented (Pearson, 2007) in various locations at larger facilities to protect screens from the damage caused by large debris.

3.1.1 Replacement of Screen Technologies

Preliminary treatment systems have a higher rate of equipment replacement than other systems at a WRRF. This results from the severe service duty within the preliminary treatment area. The incoming corrosiveness of the wastewater, the physical stresses that repeating high-flow events and their associated high levels of debris and grit can incur, the humidity around the screening system, if placed inside, or the inherent decreased longevity of the equipment, if placed outside, are all factors that lead to preliminary treatment systems having a shorter service life than other equipment at a WRRF. This offers a perfect opportunity to change to a newer screen technology and upgrade the screen system and its ancillary equipment more frequently to optimize the performance of the screening process vs the demands required by wet weather flows.

For facilities with articulated rake-type screens that inherently have a long cycle time, installation of a multiple-rake screen will provide faster cleaning. The faster cleaning may allow the facility to consider installing screens with a smaller aperture size. For facilities that have screens with slow cleaning rates or facilities with screens that inherently have higher carryover percentages, such as element belt screens, these screens can be replaced with newer, more efficient technology.

3.1.2 Addition of Screens

Where additional wet weather flow is contemplated for treatment at the facility, adding screens in parallel channels provides the most secure solution. Sufficient screens in parallel are added to process the peak wet weather flow and are taken online and offline as necessary to match target approach velocities and headloss. This will maximize treatment capacity, limit risk of sediment deposition, and minimize wear on machinery. Another option that can be considered is the construction of coarse screens for use during wet weather flows and isolatable during normal flows.

As described earlier, installation of sequenced screens may be contemplated. The first set of screens provides first-stage screening with a coarser screen and a second stage provides further screening with a finer screen. This is commonly contemplated in three instances. The first is with facilities that were initially designed with the screen process following the grit process. The second is with facilities that receive a high fraction of flow from a combined sewer that does not have independent screening and thus may send to the facility large and destructive debris. The third is when a secondary process is being installed that requires microscreening; however, this finer screen application is often performed between primary treatment and secondary treatment and, therefore, is not part of the preliminary treatment flow scheme. Sequenced screening should be avoided, where possible, because the facility is essentially provided with twice the equipment to operate and maintain.

3.1.3 Ancillary Equipment

The larger debris that can be part of the wet weather flow can jam washer/compactors. This can also occur with the simple overloading of the washer/compactors resulting from the first flush load. It is important that the washer/compactor system be conservatively sized to process the peak instantaneous debris load that it may see. Often, systems are designed such that the screens can remove the debris, but cannot adequately process the captured debris. Conservatism is important in the design of the washer/compactor system. The use of large, centralized washer/compactors fed from sluices can aid in overall performance as uneven debris, which frequently accompanies wet weather events; loads onto screens can be more readily balanced with this type of system. Washer/compacters can also be provided on rollers such that a standby unit can be placed in the area as an emergency backup. Sequenced screening systems help in this regard as the installation of coarser screens, which discharge to conveyor belts or directly into storage bins, protect screening systems that use washer/compactors. This larger debris can jam washer/compactors and other machinery such as screw conveyors (when used, although they are not recommended) at WRRFs serving combined sewer systems.

There are three typical conveying systems: belt conveyors, screw conveyors, and sluices. Each can experience degraded performance resulting from the increased load caused by a wet weather event. Screw conveyors should be avoided if larger debris or large slug loads of debris will be removed during the first flush. Sluices may also become bound with larger debris; however, this debris is easier to remove from a sluice than from a bound screw conveyor. Sluices also tend to provide the cleanest method of transporting screenings to their washing/compaction point (Mihm et al., 2009). Transport of sluicings directly into the compactor is preferable, either directly if on the same floor or via a drop box if on a lower floor. If transporting sluicings further, the use of a wet well with grinder and nonclog pumps or grinder pumps can be contemplated; however, great care during design is required to avoid buildup in the well and pump and pipe clogging. Belt conveyors are typically only used with screens that have aperture sizes greater than 10 mm and never with screens that use spray water to supplement the screen element cleaning process.

Many facilities have contracted screenings pickup and only larger facilities handle their own screenings storage bins. Depending on the size of the facility, this contract removal may be daily, periodically through the week, or weekly. The design of the storage system should be for peak screenings collected over the contracted pickup/removal time. This peak will almost certainly fall during wet weather events and especially during the first significant wet weather event of the season.

Spray water, if required, should be designed to be of a high quality to minimize the risk of clogging spray nozzles in the system and should be designed with a straining system smaller than or equal to the nozzle orifice of the spray water system. At high flow, when cleaning the screening elements is imperative, it is critical that the wash water system function properly and be appropriately sized to provide the flow and pressure required for all screens and all other preliminary treatment processes that may require supplemental water, such as washer/compactors and sluices.

System redundancy and quick-install standby equipment are critical design considerations because equipment redundancy can be difficult in screening systems. For example, conveyors and sluices provide unique redundancy issues. Typically, redundant conveyors and sluices are not provided. Additionally, many screens are coupled with dedicated washer/compactors, effectively doubling the mechanical risk of failure for a single system. Staggering screens is a method to add conveyor or sluice redundancy and using sluices for multiple parallel washer/compactors is a method to add washer/compactor redundancy. Sluices are nonmechanical and, much like pipes, are typically not provided with redundancy. When designed properly, a sluice will only infrequently experience clogging or other types of downtime.

3.1.4 Bypassing

Optimally, a bypass will be passive and will also be screened. Manual coarse screens are the most common, but can quickly accumulate debris. When in use, they require close monitoring and frequent manual raking. Bypass channels and bypass piping with motor-operated gates or valves set to open on high level are less optimal than passive systems, but are often necessary depending on the original facility configuration. Careful hydraulic analysis of passive bypassing is required because the flow that can be bypassed by a passive system can be substantially less than what a designer may intuitively feel will pass.

3.1.5 Storm Screens

Most screen types are described in *Design of Municipal Wastewater Treatment Plants* (WEF et al., 2010); however, storm screens were not contemplated in the manual and, as such, a brief overview of them is provided here. These screens can be used within the collection system or within the WRRF to overflow screened effluent to equalization to avoid deposition of debris within the equalization basin or other receiving structures.

3.1.5.1 Rectangular Deflection Storm Screen

The rectangular deflection screen is placed vertically or horizontally between the main wastewater channel and the overflow channel to receive screened wastewater. Screened flow passes through the screen, while debris is captured on the screen face. A combing/raking mechanism passes along the screen to prevent blinding and keeps debris in the flow directed toward the final debris removal point, typically located at the preliminary treatment process of the WRRF or at a screening process at a downstream pumping station. The aperture size of these screens is typically 4 mm and screens vary in length and height and can be stacked to customize the machinery for the particular application. The combing/raking devices are either hydraulically driven with their support equipment located above the water surface and within an area suitable for periodic maintenance or with submersible motors. Figures 11.3 and 11.4 provide a photograph and rendering, respectively, of a rectangular deflection storm screen.

3.1.5.2 Cylindrical Deflection Storm Screen

The cylindrical storm screen functions similar to the rectangular deflection storm screen; however, it consists of a perforated half cylinder with a rotating screw within. Screened flow passes through the perforated portion, while debris is captured on the screen face. The rotating screw includes brushes that, when rotating, clean the screen to prevent blinding and direct the debris onward in the original flow path of the wastewater. The aperture size of the perforations is typically 6 mm; however, they can be obtained in

FIGURE 11.3 Rectangular deflection storm screen (courtesy of WesTech Inc.).

FIGURE 11.4 Rendering of a rectangular deflection storm screen (courtesy of WesTech Inc.).

other sizes. The half cylinders vary in length and diameter and can be placed parallel, depending on the need of the application, or inverted, depending on the design professional's preference. Figures 11.5 and 11.6 provide two photographs and a rendering of a circular deflection storm screen.

FIGURE 11.5 Cylindrical deflection storm screens (courtesy of Huber Technology).

FIGURE 11.6 Rendering of a cylindrical deflection storm screen (courtesy of Huber Technology).

3.2 Grit Removal

3.2.1 Number of Grit Removal Units

The number of grit removal units may be limited by site constraints; however, placing additional grit removal units in service as flows increase is an option for managing wet weather flows where site constraints do not limit the number of units.

Performance vs redundancy needs to be taken into consideration. If the hydraulic capacity can be passed through the grit removal system at a reduced grit capture rate with one unit out of service, is that sufficient? Is there primary sludge degritting, reducing the need for efficient grit removal? This becomes a policy/risk management issue that requires input from the owner. It may be more cost-effective to degrit primary sludge rather than expand grit removal facilities to handle peak wet weather flows.

3.2.1.1 Aerated Grit

Aerated grit removal units can probably handle the widest range of flows of any of the grit removal options with minimal hydraulic effects. These units also have the potential for chemical addition upstream of chemically enhanced primary treatment. Units may be taken offline under average conditions to conserve energy; however, no adverse process effect would occur if all units are left in service.

3.2.1.2 Vortex—Mechanically Enhanced

Approach velocity is critical for these units and should be reviewed with the equipment manufacturer. One manufacturer requires a velocity between 0.5 and 1.1 m/sec (1.6 and 3.5 ft/sec) for its grit removal unit. The level in the grit removal unit is controlled using a weir, proprietary baffle, or some other device to maintain the water surface in the grit removal unit and the inlet velocity within the manufacturer's stated range. If the water surface elevation associated with maximum inlet velocity is observed, additional units should be placed in service.

3.2.1.3 Vortex—Tangential with Multiple Trays

The primary issue with this type of unit is the inlet headloss under wet weather flows. Multiple units are required to handle a wide range of flow conditions and stay within the manufacturer's rated flow and to limit the headloss across the unit.

3.2.2 Passive Bypass/Overflow Weir

If site constraints limit the number of grit removal units on the site, a means of bypassing a portion of the wet weather flow around the grit removal units directly to primary treatment should be considered. This could be integrated with chemically enhanced primary treatment and/or degritting of primary sludge.

3.2.2.1 Combined Sewer Overflow Screens

Consider bypassing all flows around screenings and grit using combined sewer overflow (CSO) screens. During a wet weather event, overflow bypassed through a CSO screen directly to primary treatment should have minimal grit.

3.2.2.2 Primary Sludge Degritting

If grit removal is bypassed, consider primary sludge degritting. Degritting of primary sludge is an option at facilities that continuously pump dilute primary sludge. For the degritting equipment to function adequately on primary sludge, the solids concentration typically needs to be less than 1.5%. Water resource recovery facilities that co-thicken primary and waste activated sludge in dissolved air floatation thickeners can degrit the bottom sludge, further enhancing grit removal. Degritting of dissolved air flotation tank bottom sludge has been done at numerous WRRFs. For facilities that do not have primary clarification to capture grit that passes through the grit removal facilities, effective grit removal is critical to prevent solids deposition in aeration basins, membrane bioreactors, and other secondary treatment units. For facilities without primary clarification, redundant grit removal units for wet weather flows are a necessity.

3.2.3 Tank Modifications

Tank modifications to both vortex-type grit removal units and aerated grit tanks should be considered to improve performance under all operating conditions.

3.2.3.1 Airflow

Older aerated grit tanks may not have the air supply tapered from the inlet to the outlet or adequate control of the airflow rate. The design airflow at the upstream end of an aerated grit tank is 28 $m^3/s \cdot m$ (5 cfm/ft) and, at the downstream end, the airflow is 17 $m^3/s \cdot m$ (3 cfm/ft). Adjustment of these values, with a value of 19 $m^3/s \cdot m$ (3.5 cfm/ft) at the upstream end and 11 $m^3/s \cdot m$ (2 cfm/ft) at the downstream end, should be provided to allow for adjustment of the airflow to optimize grit capture. Airflow modulation ability to increase airflow during increased flows and lower airflows during lower flows will provide the facility with the ability to target grit removal during wet weather events and minimize organic capture during normal flow periods.

3.2.3.2 Baffles

Short-circuiting of grit through aerated grit tanks has been observed. Baffles may be added to reduce the amount of short-circuiting. The use of computational fluid dynamics and/or construction of a physical model should be considered to optimize baffle configurations and locations (McNamera et al., 2012).

3.2.4 Ancillary Systems

3.2.4.1 Grit Pumps and Piping

Grit pumps should be the recessed impeller severe duty type. Suction piping should be flooded and kept as short as possible. Grit piping should be at least 102 mm (4 in.) in diameter, resulting in a minimum pump capacity of 890 L/min (235 gpm) to maintain a velocity above 1.8 m/sec (6.0 ft/sec) in the piping. For wet weather facilities, larger pumps with a capacity of approximately 2000-L/min (530-gpm) and 152-mm (6-in.) diameter piping should be considered. Pumps should be constant speed to maintain a suitable velocity in the grit piping. If possible, provide two pumps per grit hopper, each with its own suction pipe. Avoid the use of cross-connections in the pump suction and discharge piping. Complex piping configurations result is a greater potential for clogging and failure of the system. Piping should be accessible for maintenance and be provided with takedown couplings to allow clogs to be cleared. As previously noted, the addition of high-pressure reuse water to the pump suction can help fluidize compacted grit. Air can also be used as an agitation source for grit hoppers and will be more effective than reuse water in that location. Both are frequently provided.

3.2.4.2 Grit Separators

Hydrocyclones and similar grit separation equipment are provided by several manufacturers. All of these units have a relatively narrow operating range and should be dedicated to a single pump. Multiple grit separators are commonly coupled to a single grit classifier. Some equipment integrates a vortex chamber as part of the unit. There is a tendency to provide headers with connections to multiple grit separators to provide redundancy. This should not be done, however, as debris in the grit will pack into tees and other fittings, clog valves, and result in total system failure. In addition, grit separators, hydrocyclones in particular, are sensitive to changes in flow. The use of a header with hydrocyclones invites a flow imbalance that will adversely affect performance. Provide a dedicated pump to each grit separator and, if redundancy is desired, provide more than one pump per grit hopper.

3.2.4.3 Grit Classifiers

Grit separators and classifiers should be selected to remove the finest particles that the grit removal unit can effectively capture, and possibly finer. This equipment, similar to the grit removal units, is typically rated based on clean grit with a specific gravity of 2.65. The actual specific gravity may be much lower because of attached organics; as such, specifying grit separators and classifiers to remove finer grit than the grit removal unit should be considered. For example, if the grit removal unit can capture 125-μ grit, it may be prudent to specify separation and dewatering equipment to target 105-μ or smaller grit. The peak loading needs to be estimated so that the classifiers will function adequately during wet weather conditions.

3.2.4.4 Grit Tank Cleaning

Consider the installation of high-pressure water cannons for cleaning out grit tanks. The minimum water pressure should be 620 kPa (90 psi). This is particularly important for aerated grit tanks.

3.2.4.5 Grit Storage

Grit classifiers typically discharge into a hopper with a grit hopper gate to facilitate removal and replacement of grit containers located below the gate or directly into a roll-off container. There are dumpster movers consisting of a rail and ram system, pivoting discharge chutes, conveyor systems, and other methods for distributing grit in roll-off containers. During a wet weather event, these storage systems may be overwhelmed, resulting in the need to discharge directly to the floor. The structure should be designed with concrete push walls so that front-end loaders can be used to clean up after an event.

3.2.4.6 Decant Tanks

For smaller WRRFs, the use of decant tanks for dewatering grit may be a consideration. These units are essentially roll-off containers with wedge wire underdrains, decant pipes, and drain pipes used to dewater concentrated grit in place of a grit classifier. For larger WRRFs with high peaking factors, decant tanks may not be viable.

3.2.5 Access to Tanks and Channels

Regardless of the precautions taken during the design of a facility, an event will occur at some time that buries a grit hopper. Access provisions for digging out grit hoppers should be provided.

Provisions should be provided for isolation of one-half of a grit tank distribution channel for cleaning. Even with an effective mixing system, grit will accumulate in channels and, at some point, will have to be cleaned out.

4.0 OPERATIONS DURING WET WEATHER

Operation during wet weather focuses primarily on bringing additional units into service or bypassing preliminary treatment systems to stay within their hydraulic capacity.

Real-time forecasting of influent flows may be used to place additional pretreatment systems in service in advance of a wet weather event. The Metropolitan Wastewater Treatment Plant in St. Paul, Minnesota, successfully implemented real-time forecasting of flows approaching the facility so that additional screens and grit removal units can be placed in service when needed. Flow meters located throughout the collection system are used to measure the flows from member communities for billing purposes. Meters serving the areas with the largest contributory flows are monitored and these values are used in a flow-forecasting algorithm to predict wet weather flows at the WRRF. The facility can then approximate future incoming flow during the wet weather event and the operations staff can proactively place additional facilities in service based on this anticipated flow.

4.1 Screening Systems

Optimally, screens are brought online as the maximum recommended approach velocity is neared and are brought offline as the approach velocity falls below the minimum desired velocity (deposition velocity). This is typically managed using the incoming flow to the facility or the level within feeding channels. An example operation strategy with four screens (three duty and one standby) would be to operate with one channel online during diurnal lows and, as flows increase, a second channel is brought online throughout the rest of the day. As flows continue to increase as a result of a wet weather event, the third and then the fourth channels are brought online. Even though the fourth channel is considered a redundant channel as long as velocities in the channels are not significantly below the settling velocity, the redundant screen channel can be put into operation. However, the system should always be capable and designed to hydraulically pass the peak hour flow with the largest screen channel offline.

Rising channel levels will also engage screen channels, more rapid cleaning cycles, and more rapid movement of the cleaning mechanisms. Rising channel levels will always govern over optimum approach velocities and often during extreme wet weather; capture performance of the screen is secondary to simply passing flow through the structure. Typically, a screen's cleaning cycle will be engaged on a timer or on high flow and many screen types have multiple speed settings such that the speed of the cleaning mechanisms increases as higher level set points are reached. During the wet weather event, the cleaning cycle will be at its highest and may even be continuous and the cleaning mechanism will be operating at their fastest speed.

During wet weather, the screens should be physically monitored by the operator much more frequently because all duty screens and, likely, all system screens will be in operation. Additionally, all ancillary systems will also be in operation. First flush grit deposition and screen binding, large debris capture, and subsequent screen or washer/compactor or conveyor jamming are more likely to occur. Screen blinding is also more likely to occur and, in facilities with bypassing to manual coarse screens, bypassing and removal of screenings from these manual coarse screens will be more likely. Whereas once to three times a shift rounds to the preliminary treatment system may be warranted during normal flow operation, nearly continuous physical operations presence is recommended during the peak flows that accompany wet weather events.

Periodic operation of the redundant screen and the screens targeted for use for high flow during typical flow is important to the successful operation of the system. Depending on the inlet design to the screen channels, debris and grit can accumulate at the face of the isolation gates. It is best to periodically operate these gates and free the accumulated debris and grit rather than to wait for the wet weather event that could then overwhelm the screen and its ancillary equipment. Proper rotation of equipment is always a key to maintaining the equipment viable when it is needed.

4.2 Grit Removal Systems

Staff of the WRRF should monitor influent flows and weather conditions to determine when additional grit removal units should be brought online. As flows increase and grit removal units approach capacity, additional units should be placed in service.

Under dry weather conditions, grit pumps are frequently run on a timed sequence to reduce energy consumption. During a wet weather event, grit pumps and associated grit separators and classifiers should run continuously to minimize the potential for burying a grit hopper or clogging of piping, grit separators, and grit classifiers. The operation of the grit pumps can be tied to an influent flow meter to switch automatically from timer-based operation to continuous-based operation during a predetermined set flow condition.

4.3 Collection System Operation to Aid Preliminary Treatment Systems

4.3.1 Pump Control

Managing the collection system to reduce grit loads during wet weather events should be considered. For WRRFs with influent pumping, use of the pumping station to draw down the interceptor to scour grit should be considered. The pumping station should have a trench-type wet well designed in accordance with ANSI/HI 9.8 (American National Standards Institute, 2012) so that the grit and floatables in the wet well are fluidized and pumped to the WRRF. The upstream effect of this practice will be limited by the drawdown curve in the interceptor and will not eliminate high grit loads during wet weather events, although it may reduce the magnitude of the grit loading. A similar procedure could be implemented throughout a collection system to scour grit from the collection system and to clean out grit that has settled at low spots in force mains.

4.3.2 Rock Traps and Grit Pits

The installation of rock traps and/or grit pits in the collection system to prevent the largest debris from reaching a WRRF has been implemented by numerous municipalities. These systems frequently fail because they rapidly fill with grit, leaving them unavailable during a wet weather event. If rock traps or grit pits are installed, routine cleaning of the facility must be done to ensure that they are available when needed. Omaha, Nebraska, has successfully implemented rock traps throughout its collection system to capture grit. About a fourth of the City of Omaha is served by a combined sewer system/collection system. Sand and gravel are applied to the streets during the winter months, resulting in large grit loads, especially during snowmelt or rain events in the early spring. The larger grit facilities are equipped with clam shells for removal of accumulated material. The smaller facilities are underground chambers that are maintained with Vactor trucks and other mobile equipment, as necessary, depending on the debris that accumulates. Odors from any of these facilities can be significant, resulting from the nature of the sediment and solids that accumulate; therefore, odor control provisions must be considered. The amount of odor control that is required is dictated by the location of the facility and in relation to the public. Omaha's current grit facilities will need to be significantly upgraded as more wet weather flows are collected during implementation of its Combined Sewer Long Term Control Plan (CSO LTCP). The experience that Omaha has gained through operation of its existing grit facilities will help the city manage the additional long-term operation and maintenance requirements as the facilities associated with its CSO LTCP come online. Other municipalities implementing long-term control plans for control of CSOs may wish to consider implementing rock traps and grit pits to mitigate grit effects at WRRFs during wet weather events.

5.0 REFERENCES

American National Standards Institute (2012) *ANSI/HI 9.8-2012: American National Standards for Rotodynamic Pumps for Pump Intake Design;* American National Standards Institute: New York.

Botero, L.; Woodley, M.; Knatz C.; Topa, T. (2011) Hydraulic Similarity of Headloss Predictions Derived Using Commonly Used Methods Versus Actual Results as it Relates to Wastewater Screen Elements. *Proceedings of the 84th Annual Water Environment Federation Technical Exhibition and Conference* [CD-ROM]; Los Angeles, California, Oct 15–19; Water Environment Federation: Alexandria, Virginia.

Casey, B.; Eimstad B.; Thompson S.; Wodrich J.; Mooney G.; Cork T.; Guthrie M. (2009) Green Screening in Bend, Oregon—Innovative Band Screen Headworks Facility Reduces Energy Consumption, Landfill Impacts and Plant Operating Costs. *Proceedings of the 82nd Annual Water Environment Federation Technical Exhibition and Conference* [CD-ROM]; Orlando, Florida, Oct 17–21; Water Environment Federation: Alexandria, Virginia.

Koch, J.; Hielema, E. (2010) Saga of Two Screens. *Proceedings of the 83rd Annual Water Environment Federation Technical Exhibition and Conference* [CD-ROM]; New Orleans, Louisiana, Oct 2–6; Water Environment Federation: Alexandria, Virginia.

McNamera, B.; Bott, C.; Hyre, M.; Kinnear, D.; Layne, J. (2012) How to Baffle a Vortex. *Proceedings of the 85th Annual Water Environment Federation Technical Exhibition and Conference* [CD-ROM]; New Orleans, Louisiana, Sept 29–Oct 3; Water Environment Federation: Alexandria, Virginia.

McNamera, B.; Layne, J.; Hyre, M.; Kinnear, D.; Bott, C. (2012) Evaluation of Three Full-scale Grit Removal Processes Using CFD Modeling. *Proceedings of the 85th Annual Water Environment Federation Technical Exhibition and Conference* [CD-ROM]; New Orleans, Louisiana, Sept 29–Oct 3; Water Environment Federation: Alexandria, Virginia.

Mihm, S.; Cronister, H. L.; Wagner, R.; Evers, M.; Gorgi E.; Irenumaagho, N. (2009) Flume Water Transport of Screenings Improves Headworks

Operations. *Proceedings of the 82nd Annual Water Environment Federation Technical Exhibition and Conference* [CD-ROM]; Orlando, Florida, Oct 17–21; Water Environment Federation: Alexandria, Virginia.

Pearson, M.; O'Kelley, S. (2007) Rock Box Protects Headworks at Kansas City, Missouri Blue River Wastewater Treatment Plant. *Proceedings of the 80th Annual Water Environment Federation Technical Exhibition and Conference* [CD-ROM]; San Diego, California, Oct 13–17; Water Environment Federation: Alexandria, Virginia.

Water Environment Federation; American Society of Civil Engineers; Environmental & Water Resources Institute (2010) *Design of Municipal Wastewater Treatment Plants,* 5th ed.; WEF Manual of Practice No. 8; ASCE Manuals and Reports on Engineering Practice No. 76; Water Environment Federation: Alexandria, Virginia.

Wodrich, J. V.; Winkler, T.; Leaf, B.; Clark, S. F.; Youker, B. (2005) Wet Weather Impacts on Influent Fine Screening System Design and Operation. *Proceedings of the 78th Annual Water Environment Federation Technical Exhibition and Conference* [CD-ROM]; Washington, D.C., Oct 9–Nov 2; Water Environment Federation: Alexandria, Virginia.

6.0 SUGGESTED READINGS

Couture, M.; Steele, A.; Bruneau, M.; Gadbois, A.; Hohman, B. (2009) Achieving Greater Efficiency for 360° Rotational Grit Removal Technology Using Empirical Data and CFD Analysis. *Proceedings of the 82nd Annual Water Environment Federation Technical Exhibition and Conference* [CD-ROM]; Orlando, Florida, Oct 17–21; Water Environment Federation: Alexandria, Virginia.

Fostner, G. (2007) Screen Selection—Understanding Your Choices. *Water Sci. Technol.,* **19** (10), 60.

Tchobanoglous, G.; Burton, F.; Stensel D. (2003) *Wastewater Engineering Treatment and Reuse,* 4th ed.; McGraw-Hill: New York.

U.K. Water Industry Research Limited (2002) *CSO Screen Efficiency (Proprietary Designs)*; U.K. Water Industry Research Limited: London, U.K.

U.S. Environmental Protection Agency (1999) *Combined Sewer Overflow Technology;* Fact Sheet, Screens, EPA-832/F-99-040; U.S. Environmental Protection Agency: Washington, D.C.

Water Environment Federation (2007) *Operation of Municipal Wastewater Treatment Plants*, 6th ed.; Manual of Practice No. 11; Water Environment Federation: Alexandria, Virginia.

12

Primary Treatment

Jurek Patoczka, Ph.D., P.E.; Henryk Melcer, Ph.D., P.E.; John R. Dening, P.E., CFM; and Mark Stirrup

1.0 INTRODUCTION

Primary treatment encompasses physical–chemical processes accomplishing separation and removal of suspended and floating solids from wastewater. The most common primary treatment process is quiescent sedimentation in primary clarifiers, with skimming devices for removal of floating matter and grease.

The chief measure of primary treatment efficiency is total suspended solids (TSS) removal. However, reduction of the organic and nutrient loads associated with the removed TSS fraction is of high interest because a critical objective of primary treatment is to reduce the loads of these constituents on the secondary system. This is beneficial because aeration basins, volume and blower capacity requirements are reduced. Lower inert and other solids loads allow operation at a higher mixed liquor volatile suspended solids fraction, thus increasing the system nitrification capacity. The phosphorus associated and settled with particulate matter will minimize phosphorus removal, whether by biological or chemical means; in the latter case instance, this reduces chemical demand.

The most common primary settling facility at water resource recovery facilities (WRRFs) is the rectangular or circular primary clarifier. Because the focus of this publication is to address the effects and management of wet weather flows on existing WRRFs, for design details of conventional clarifiers, the reader is referred to *Clarifier Design* (WEF, 2005) and *Design of Municipal Wastewater Treatment Plants* (WEF et al., 2010). Application of plate and tube settlers (Lamella clarifiers) for primary treatment is limited in the United States and Canada; however, because they are of potential interest for wet weather flow management, they are discussed in the following section. Stacked clarifiers are uncommon in the United States and Canada, with the exception of the Deer Island facility in Boston, Massachusetts.

Alternative primary or equivalent treatment concepts include chemically enhanced primary treatment (CEPT) or chemically enhanced settling (CES), ballasted flocculation, vortex separators, fine-screen (or micro-) filtration, and high-rate filtration using novel media; these concepts are discussed either later in this chapter or in Chapter 14.

2.0 EFFECTS OF WET WEATHER FLOWS ON PRIMARY TREATMENT

The primary clarification process can be adversely affected by wet weather flows in the following ways:

- High solids loading in the first flush, causing high sludge blanket levels;
- Scouring of solids from the sludge blanket, resulting in excessive solids in the primary effluent;
- Reduction in overall removal efficiency of biochemical oxygen demand (BOD) and TSS, resulting from elevated surface overflow rates (SORs);
- Excess grit and screenings loadings to primary clarifiers resulting from overloaded preliminary treatment processes; and
- Flooded scum removal and storage boxes.

These are discussed in the following subsections.

2.1 Effect on Primary Clarifier Removal Efficiency

2.1.1 Effect of Overflow Rate

Overflow rate is considered to be the primary design parameter affecting performance of the sedimentation process. Figure 12.1 presents historically used relationships (U.S. EPA, 1978) between overflow rate and TSS and BOD removal efficiencies. The TSS removal efficiencies, typically in the 45 to 65% range for the design (average) conditions, decrease to 30 to 45% at peak hourly flows. For BOD, the corresponding guidance is 25 to 35% at the design flows, with 15 to 25% at the peak flows.

Although the guidance presented in Figure 12.1 should, in general, be valid for removal efficiency of the influent with the same characteristics, data from full-scale facilities frequently reveal only a weak correlation between the overflow rate and TSS removal efficiency (Wahlberg et al., 1997). This

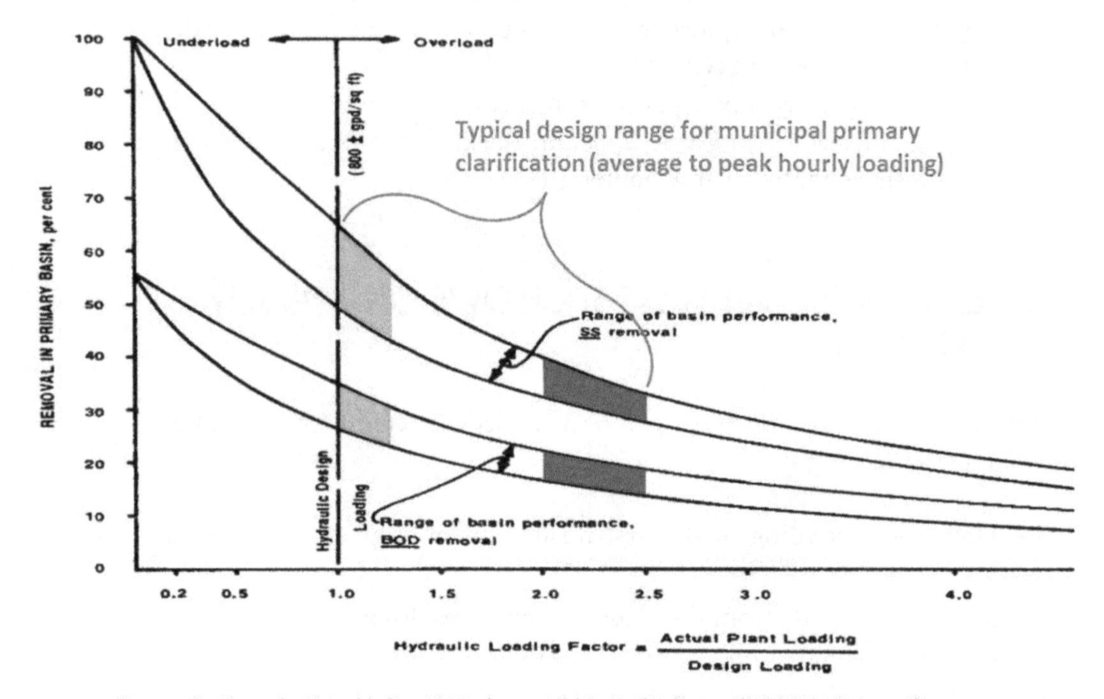

Source: Shading and notes added to U.S. Environmental Protection Agency (1978) *Field Manual for Performance, Evaluation and Troubleshooting at Municipal Wastewater Treatment Facilities*, EPA 430/9-78-001, Figure 17

FIGURE 12.1 Traditional guidance showing the relationship between overflow rate and primary clarifier performance in terms of TSS and BOD removal efficiency. A U.S. EPA-recommended (at that time) design overflow rate of 32.6 m/d (800 gpd/sq ft) is shown on this graph; in practice, most primary clarifiers will operate at overflow rates lower than this during dry weather operation (Fitzpatrick et al., 2012).

is because influent characteristics, including raw wastewater TSS concentration, particle distribution including nonsettleable fraction, flocculating properties, and temperature can have a significant effect on the TSS removal efficiency. These factors and state-of-the-art modeling of primary clarifier performance are discussed in detail in *Design of Municipal Wastewater Treatment Plants* (Chapter 12; WEF et al., 2010) and Chapter 2 of *Clarifier Design* (WEF, 2005).

It is particularly important to recognize that quiescent settling (i.e., primary clarification unaided by chemical addition) is capable of removing only this fraction of TSS that is settleable. Nonsettleable TSS consist of fine and colloidal matter and are operationally defined as the fraction of the TSS that does not settle upon 30 minutes of flocculation (w/o chemicals) followed by 30 minutes of settling (WEF et al., 2010). The settleable fraction varies significantly, particularly in response to wet weather events, making performance prediction and modeling challenging.

Chemical addition in the form of coagulant and/or polymer, dramatically increases the proportion of settleable solids, allowing for a significant increase in TSS removal efficiency at the same overflow rate or, more importantly, from the perspective of this publication, facilitates adequate TSS removal at significantly higher SOR. The CEPT is discussed in more detail later in this chapter.

In addition to nonsettleable TSS, other characteristics of the raw wastewater change significantly, as discussed in more detail in Section 3.0. It was observed (as summarized in, Chapter 3 of *Clarifier Design* [WEF, 2005]) that solids present in the wet weather flows are typically of larger size and better settleability (grit-like material scoured in the initial phases of a wet weather event) than dry weather influent. Dramatic variability in influent solids settleability during wet weather events undoubtedly contributes to the typically weak relationship between overflow rate and performance.

In summary, it could be stated that, on average, as the overflow rate increases, percent removal efficiency will tend to decrease, although a considerable variability in the actual performance could be expected resulting from a number of site-specific and event-specific factors.

In assessing the effect of primary clarification overload, or inferior performance during wet weather events, it should be recognized that primary treatment at a WRRF is primarily there to lower loadings of pollutants of concern on the downstream processes (and many WRRFs do not have primary clarification at all). Consequently, inferior performance of primary clarification during wet weather events is of concern only as far as it causes problems at the downstream processes (which is typical). A significant exception is when part of the primary effluent is diverted around the secondary treatment, and its quality could directly affect the receiving waterbody.

2.1.2 Effects of Temperature, Total Dissolved Solids, and Other Parameters

As discussed in Chapter 9, wet weather flows are typically colder, which increases water viscosity and thus lowers settling velocity of particulate matter. Wet weather flows, particularly the first flush, could also have a higher concentration of total dissolved solids (TDS), particularly if deicing salts were in use. Both temperature and TDS concentration could cause the specific density of the incoming wastewater to be different than the clarifier contents, causing density currents resulting in a poorer performance. On a related note, windy conditions frequently accompanying wet weather events could themselves cause unwanted circulation of the clarifier contents. Effects of changes in other wastewater characteristics, such as pH, are difficult to quantify.

2.2 Effect on Primary Fermentation

At some enhanced biological nutrient removal (EBNR) facilities, primary clarifiers are used as prefermenters or "activated primary sedimentation tank", a practice originally described by Barnard et al. (1984). The additional volatile fatty acids (VFAs) generated in the activated primary tanks by a slow recirculation of the settled sludge back to the clarifier influent are directed to the anaerobic and/or to anoxic zones to enhance the rate of biological phosphorus removal and denitrification, respectively. At the wet weather flow conditions, the efficiency of the VFA production by primary clarifiers could decrease significantly. This results from the higher rate of primary solids accumulation, which necessitates a higher sludge withdrawal rate to prevent solids buildup (and washout) from the primary clarifier. Additionally, the rate of fermentation will be negatively affected by lower raw wastewater temperatures typically associated with wet weather flow. Finally, the concentration of VFAs in the primary effluent will be lowered because of the dilution. All these factors will decrease the supply of VFAs to the downstream EBNR processes, exacerbating a negative effect of wet weather flows on these processes.

2.3 Floatables Control

Wet weather events cause an increased load of floatables reaching the head of the facility, particularly during the first flush conditions. These include grease and scum and particulate floating matter that was not removed during preliminary treatment. Most of the preliminary systems (bar screens) are equipped with diversion features, which could be activated during severe flow conditions, significantly increasing quantities of floating debris reaching the primary clarifiers. Scum baffles protecting effluent weirs should be continuous and of adequate depth to prevent floatables from escaping the clarifier.

The typical scum and floatables control devices are a scum box with a beach plate (Figure 12.2), common in circular clarifiers, and rotating (or tilting) scum troughs, which are used mostly in rectangular clarifiers (see Figure 12.3). Other mechanisms such as a paddle wheel or a telescopic valve are less common. Where manual activation of rotating scum troughs is practiced (or override is available), these devices may have to be activated with an increased frequency at high flows.

2.4 Sludge Generation and Handling

Solids loading on the primary clarifiers will increase significantly during wet weather flows, particularly resulting from the first flush effect in initial

FIGURE 12.2 Scum box and beach plate at a small primary clarifier (courtesy of Monroe Environmental, Monroe, Michigan).

hours (see Chapter 2). This will lead to accelerated accumulation of solids in the clarifier, so deeper clarifiers (i.e., clarifiers with longer high-rate treatment [HRT], SORs being equal) will be better able to accommodate such loadings and disengage the sludge blanket from hydraulic currents that will occur at higher flows. The rate of primary sludge withdrawal should be increased to prevent elevated blanket levels and scouring of the solids. Higher overflow rates and less opportunity for thickening will result in more dilute sludge generated at a higher volumetric rate. This can overtax downstream thickening and sludge processing facilities, particularly anaerobic digesters, if adequate thickening and/or storage facilities are not available (see Chapter 16).

During the wet weather events, preliminary treatment facilities may be operating at the upper limits of their capacities (see Chapter 11), resulting in grit carryover and increasing primary sludge generation. If sludge degritting is not practiced, anaerobic digesters (if used) will be burdened with extra inert solids. Primary sludge yield and composition during wet weather events is discussed in Chapter 16.

FIGURE 12.3 Rotating scum trough (courtesy of Jim Myers & Sons, Inc., Charlotte, North Carolina).

2.5 Effects on Downstream Processes

As discussed previously, the main function of primary clarifiers is reduction of pollutant loading on the secondary process and retrieval of degradable organic matter for energy recovery through anaerobic digestion. During the early part of a wet weather event, higher loadings and lower primary clarifier efficiency will increase the solids and organic loading on the secondary system, resulting (in activated sludge systems) in an initially accelerated generation of mixed liquor solids and a change in their composition (see Chapters 13 and 16).

Higher flows will have a pronounced effect on secondary clarifiers, potentially requiring a range of measures such as conversion to a step-feed or contact stabilization mode or activation of polymer addition, as discussed in Chapter 13. Wet weather events could cause nitrifier washout and nitrification failure. Reduced hydraulic retention time and, frequently, lower temperatures associated with wet weather events could exacerbate problems maintaining nitrification, denitrification, and biological phosphorus removal, as discussed in more detail in Chapter 13.

2.6 Effects of Other Processes

2.6.1 Co-Settling with Waste Biosolids

In older WRRFs using tricking filter or rotating biological contactor fixed film systems, a common practice was to direct waste sludge from the final clarifiers to the primary clarifiers for co-settling. Wet weather flows will significantly increase sloughing of biomass from fixed film processes caused by hydraulic shear. If this elevated solids load were transferred to the primary clarifiers at the time of high flow, it would exacerbate hydraulic and solids load stress on the clarifiers and could contribute to solids loss from the primary clarifier. If practical, waste solids from fixed film processes should be stored in the final clarifier or directed to any other available storage facility during wet weather events. Although co-thickening of waste activated sludge is less common, a similar strategy should be applied to those facilities.

2.6.2 Effect of Backwashes and Other Sidestreams

Similar consideration should be given to other return streams, such as filter backwashes or streams from sludge processing facilities. Filtration facilities will likely be heavily taxed during wet weather flow conditions both hydraulically and because of the elevated TSS concentration. If possible, frequency of backwash should be controlled to a lowest level practical by tolerating a higher pressure loss or adjusting timer settings to minimize additional flows to the primary clarifier.

3.0 OPTIONS FOR WET WEATHER FLOW MANAGEMENT AND TREATMENT

3.1 Additional Primary Treatment Capacity

3.1.1 Additional Primary Clarifiers

Where possible, a straightforward resolution of limited wet weather primary clarification capacity, particularly when peak to average flow ratios are greater than 5:1, is the deployment of additional or enhanced sedimentation facilities (Fitzpatrick et al., 2008). This may be feasible at older facilities with abandoned primary clarifiers or similar structures or where some clarifiers are offline during typical flow conditions resulting from adequate capacity. An example includes the city of Auburn, New York, where an old, abandoned primary treatment facility was retrofitted to accept and treat excess wet weather flows (U.S. EPA [2000]; see also the case study in Section 5.2). Provisions for emptying or flushing intermittently used tanks after wet weather events should be considered (Leffler and Harrington, 2001). However, purposeful construction of additional primary clarifiers for wet weather flow treatment is unlikely to be economical given the availability of other processes and approaches discussed in subsequent sections. In particular, coagulant and polymer addition to the existing primary clarifiers (CEPT) has been demonstrated to be an effective approach, as detailed in Sections 3.2 and 4.3.

3.1.2 Lamella Settlers

Plates or tubes installed at an angle in a clarifier (or a part of it) will significantly increase the effective settling area available within the same footprint. Additionally, short settling distances promote close contact and flocculation of solids. Such an arrangement, commonly known under the trade name Lamella (Parkson Corporation, Fort Lauderdale, Florida), is frequently used in potable water treatment. Application in wastewater treatment is limited by concerns about solids, floatables, and oil and grease buildup and/or fouling by biological growth, although more than 100 applications in primary (and a few in secondary) clarification are known (WEF, 2005). Almost all of these are in Western Europe, primarily in France, with a few in Canada and none in the United States. However, many ancillary or tertiary wastewater treatment processes in use in the United States, such as Actiflo or Densadeg, use Lamella inserts in the clarification section of their treatment train.

The basic configuration typically consists of rows of inclined, parallel plates or crossing plates forming bundles of tubes. They are typically installed at the clarifier surface at a depth of up to 2 m. Various flow patterns are being used, with the upflow-cross-flow pattern appearing to offer the best ability to separate both settleable and floating matter. Fine screening and good grit and oil and grease removal before the Lamella is important for successful operation. In wastewater applications, provisions for easy access to the Lamellas for maintenance and flushing is critical. Figure 12.4 illustrates a Lamella application offered for primary treatment in Europe.

The overflow rates achievable by a Lamella may be up to 10 times higher than for conventional clarification and effective overflow rates of up to 15 m/h (8800 gpd/sq ft) were reported. This is a result of the effect of the combined, vertically projected surface area of the Lamella plates/tubes. However, the space savings offered by these devices are significantly less than theoretically calculated from the effective surface area because additional footprint is necessary to accommodate additional flow distribution and inlet and outlet structures. In addition, Lamella packs are frequently installed only in a part of the overall tankage, as illustrated in Figure 12.4. For details concerning Lamella modeling and design calculations, the reader is referred to *Clarifier Design* (WEF, 2005).

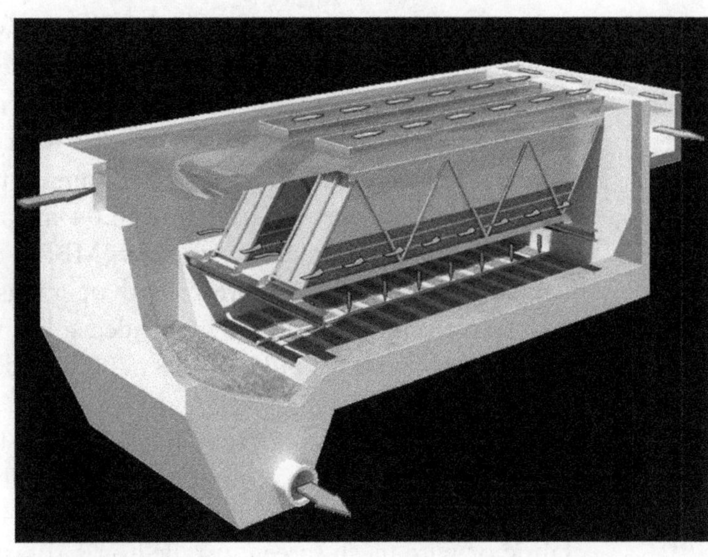

FIGURE 12.4 Plate- and tube-type settlers (courtesy of Hydro International, Ely, Cambridgeshire, U.K.).

Chemical addition ahead of Lamella plates will improve solids separation, just as it would be the case with conventional settling. Tests performed at King County, Washington, facilities demonstrated that the use of plates greatly increased the already impressive allowable SOR for CEPT treatment, from 8.5 to 34 m/h (5000 to 20 000 gpd/sq ft) (Crow et al., 2012). The acceptable performance was defined here as a TSS removal efficiency better than 50% on a consistent basis.

3.1.3 Swirl Concentrators

Swirl concentrators, also known as *vortex solids separators*, originate and are primarily used for stormwater and combined sewer overflow (CSO) treatment, although they could be incorporated to the conventional wastewater treatment train either as the main-line or bypass process.

Vortex separators accomplish separation of grit and readily settleable solids by inducing centrifugal motion of wastewater in a cylindrical vessel with tangential inlet structure. Some vortex separators include floatables removal and serve as flow regulators for CSO application. Chemical addition could improve their performance in terms of TSS removal.

3.1.4 Fine Screens

There is a continuum between coarse or bar screens used for preliminary treatment (as covered in Chapter 11) and fine screens and microscreens used for an increasing TSS removal in raw wastewater. Fine screens are typically used to remove material that may create operational and maintenance problems in downstream processes, particularly in systems that lack primary treatment. Recently, screen sizes with openings of 6 mm and even 3 mm are becoming standard for preliminary treatment (Chapter 11). Fine screens with an opening of 1 mm (0.04 in.) are commonly deployed for protection of membrane bioreactors (MBRs) and would typically require two-stage screening. Fine screens with openings below 1 mm are technically microscreens and can reduce suspended solids to levels near those achieved by primary clarification (U.S. EPA, 2003) and, as such, are discussed here in more detail.

Conventional microscreens are commonly used for polishing of secondary treatment effluent to eliminate biological flocs from escaping a clarifier, although their popularity in this application is declining. They consist of a fabric or wire mesh screen installed on a rotating drum.

Recently, microscreens designed for filtering raw wastewater have been introduced. Figures 12.5 and 12.6 show schematics and photograph, respectively, of the M2R microscreen (M2 Renewables, Lake Forest, California). Screened and degritted wastewater is filtered through a continuous belt screen made of polyester. The removal efficiency claimed by the manufacturer is comparable to primary clarification (40 to 70% TSS removal), although this is expected to be a function of the mesh size (belts with openings down to 105 µm were tested). The screen filters out solids, which are scraped from the mesh, dewatered with an auger screw, and discharged as cake with 30 to 40% dry solids. A backwash system is used to prevent clogging of the screen, with availability of hot water recommended to aid in the removal of oil and grease. The M2R microscreens have been piloted at several locations and are reported to be used at several industrial sites.

Another provider of emerging equipment of similar construction is Salsnes Filter AS of Namsos, Norway (represented in North America by Trojan Technologies). This construction uses fine mesh with 100- to 500-µm openings attached to an inclined rotating wire cloth belt (U.S. EPA, 2013).

The appeal of microscreen technology for wet weather primary treatment is that it requires much less surface area and it is claimed to cost less than primary clarifiers with equivalent capacity (U.S. EPA, 2013). However, long-term, continuous operating experience at municipal WRRFs is needed to assess wider application of this technology for raw wastewater treatment, taking into account maintenance requirements and odor control implications.

Advancements in high-rate filtration technologies, such as compressible, cloth, and upflow floating media filtration, as applied to wet weather flow treatment, are discussed in Chapter 14.

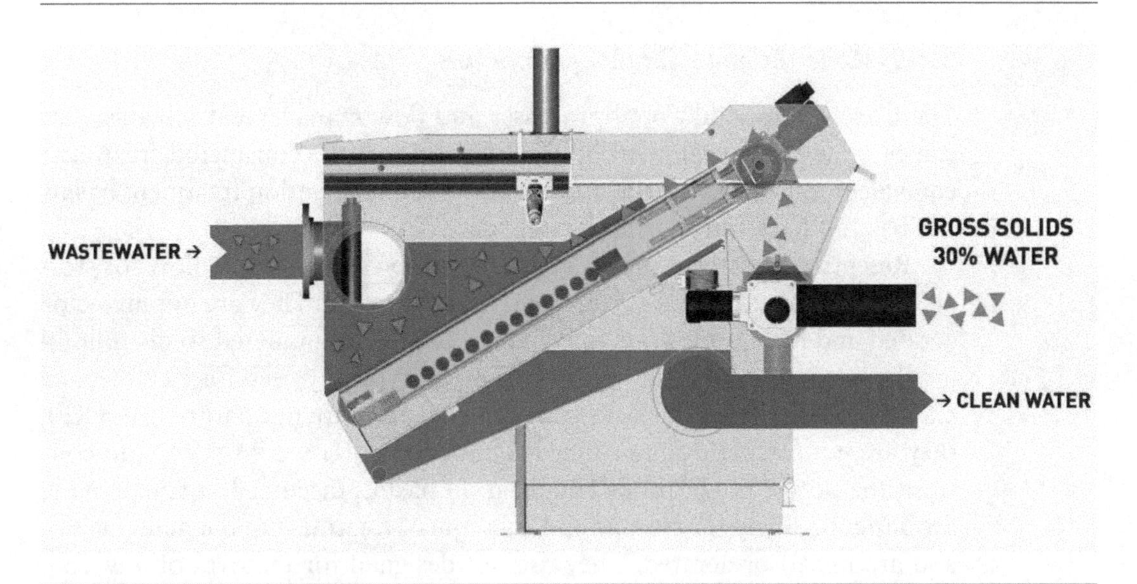

FIGURE 12.5 Schematics of a M2R microscreen for raw wastewater treatment (courtesy of M2 Renewables, Lake Forest, California).

FIGURE 12.6 A photograph of a M2R microscreen for raw wastewater treatment (courtesy of M2 Renewables, Lake Forest, California).

3.1.5 Other Primary Treatment Concepts

There are several additional processes and flow management schemes that can provide partial removal of TSS and thus can be considered treatment equivalent to primary clarification. These include retention treatment basins (RTBs) and high-rate clarifier systems.

Retention treatment basins are designed to capture excess flow for storage and partial treatment for TSS and/or disinfection. They are not mixed or aerated and have provisions for removal of any accumulated solids following the wet weather event. Because RTBs are typically used for treatment of CSO or stormwater at remote locations (i.e., they are not part of a WRRF), they are not further discussed in this publication (U.S. EPA, 1999). In contrast, the offline retention or equalization basins, discussed in Chapter 10, are designed solely for storage and subsequent treatment in the main facility and are mixed or aerated. They are not designed for removal of any TSS, although, in practice, some settling is unavoidable and such flow storage and equalization facilities are typically equipped with sludge removal and cleaning provisions, as discussed in Chapter 10.

There are various treatment processes, primarily for removal of TSS, which enhance the conventional gravitational settling by various chemical and mechanical means and are typically referred to as *HRT* or *enhanced high-rate treatment* (*EHRT*) processes. Chemically enhanced primary treatment is discussed in detail in the subsequent section. Other HRT (EHRT) processes include various forms of ballasted flocculation and filtration with novel media. The original target application for these processes was frequently treatment of CSOs and stormwater, although they could be incorporated to the conventional wastewater treatment train either as the mainline or side-stream process. The aforementioned vortex separators are also sometimes classified as HRT processes, although their performance could be inferior to other HRT processes (hence, introduction of the designation EHRT for high-rate, high-performance processes). Design and application of HRT (EHRT) processes at WRRFs is discussed in more detail in Chapter 14.

3.2 Chemically Enhanced Primary Treatment

Chemically enhanced primary treatment is useful for facilities that have already invested in primary clarifier capacity. Capacity can be increased by a factor of 2 to 4 depending on the peaking factors associated with wet weather flow conditions and the hydraulic constraints of existing primary facilities. Wet weather SORs can be elevated to as high as 9.3 m/h (5500 gpd/sq ft) from the typical dry weather SORs of 1.7 m/h (1000 gpd/ sq ft) or less. Figure 12.7 shows a comparison of TSS removal performance vs SOR in CEPT systems vs conventional primary clarifiers. Chemically enhanced

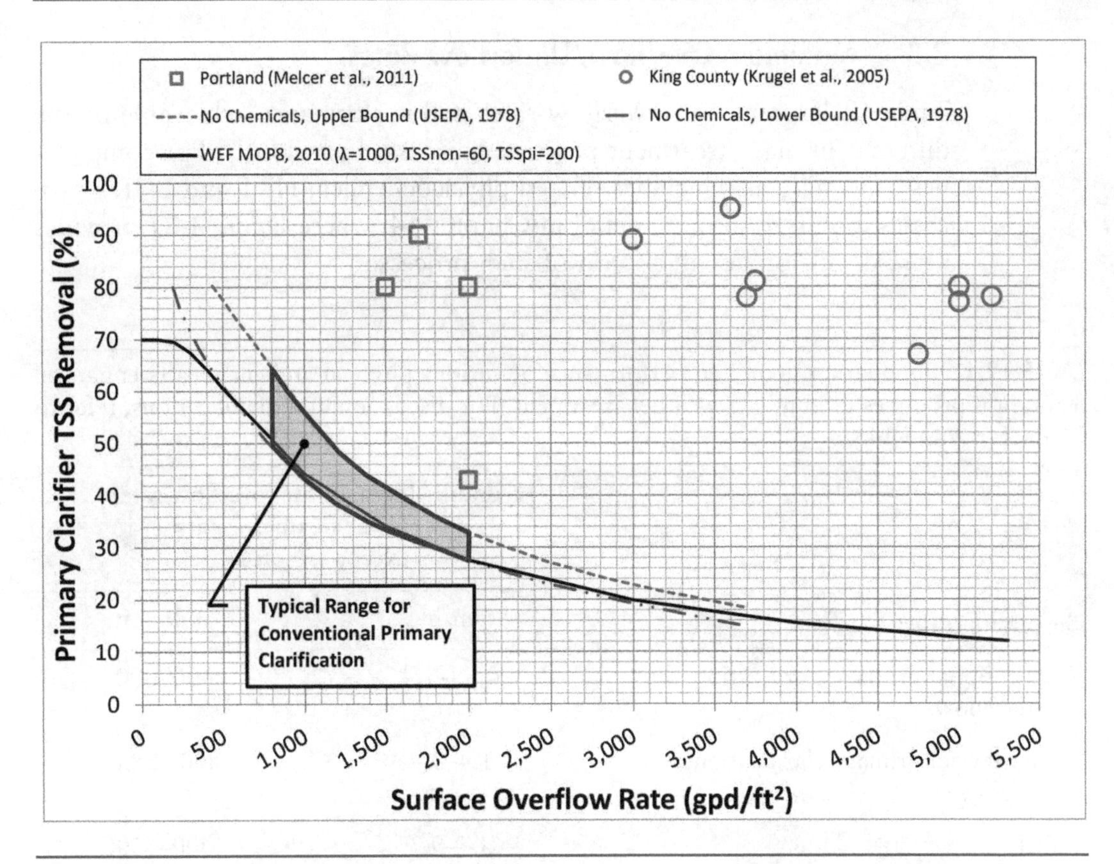

FIGURE 12.7 Comparison of CEPT systems with conventional primary clarifiers.

primary treatment data are drawn from the two case histories in this chapter and, for comparison, this figure also shows conventional sedimentation curves from U.S. EPA's *Field Manual* (1978) along with a theoretical performance curve based on the equations found in Chapter 12 of *Design of Municipal Wastewater Treatment Plants* (WEF et al., 2010). As illustrated by this figure, CEPT has significantly higher performance capabilities than gravity settling alone and can typically be expected to double or triple the capacity of a primary clarifier.

If these constraints can be addressed at a reasonable cost, then CEPT is a more cost-effective solution to managing wet weather flows than constructing new high-rate clarification facilities. An alternative way of implementing CEPT with new primary clarifiers is to consider designing them to be operated in a dual-use mode. At low flow, they are operated without chemicals; as flows increase, each clarifier is progressively converted to operating with chemicals. This was done effectively at King County, Washington's, greenfield Brightwater facility in the Seattle metropolitan area (Krugel et al., 2005).

3.3 Maximum Overflow/Underflow Rates

Table 12.1 summarizes overflow rates and performance achievable by the different primary treatment processes discussed previously. For completeness, the table also includes related physicochemical HRT and EHRT processes discussed in more detail in Chapter 14, where Figure 14.1 provides more details on design criteria of such processes.

TABLE 12.1 Summary of expected performance of various physical and chemically enhanced separation processes used in wet weather treatment. For discussion and qualifications, refer to Chapters 12 and 14.

	Hydraulic loading rate (overflow rate) during wet weather events (except as noted)	
Separation process	m/h	gpd/sq ft
Conventional primary clarification, dry weather flow	0.68–1.4	400–800
Conventional primary clarification, wet weather events	1.4–3.4	800–2000
CEPT	3.4–9.3	2000–5500
Plate or tube settlers (Lamella)	up to 15	up to 8800
Vortex separators (w/o chemicals)	4–10*	2400–5900*
Vortex separator (w/ chemicals)	4–40*	2400–24 000*
Microscreens	N/A	
Plate or tube settlers (Lamella) with chemicals (CES)	Refer to Table 14.1	
Dense (recirculated) sludge		
External ballast		
Compressible media filtration		
Cloth media filtration		
Floating media filtration		

*WEF, 2006

4.0 DESIGN AND MODIFICATION CONSIDERATIONS

4.1 Regulatory Considerations

The regulatory framework for operation of WRRFs at wet weather flows is provided in Chapter 2. As discussed there, the practice of diverting part of the wet weather flow around secondary treatment is controversial, but it remains a fact of life for many facilities compelled to protect biological treatment processes. In such instances, primary sedimentation may be the only treatment (apart from disinfection) provided for part, or all, of the flow reaching a WRRF during a wet weather event. Therefore, adequate performance of primary treatment under such circumstances is critical for meeting the relevant permit limits. As discussed in Chapter 2, lower influent concentrations of TSS and 5-day BOD during the wet weather events could exacerbate any percent removal requirements.

4.2 Primary Clarifiers

As a standard design practice, primary clarifiers should be designed to perform adequately under peak wet weather flow conditions identified in the planning phase (see Chapter 2). Refer to appropriate Manuals of Practice (i.e., WEF [2005] and WEF et al. [2010]) for detailed design guidance and to Section 2.1.1 and Table 12.1 of this chapter for a discussion of primary clarifier performance at high flowrates. Chapter 5 provides a discussion on hydraulic considerations relevant to flow distribution and inlet and outlet structures of primary treatment facilities.

4.2.1 Stress Testing

Water resource recovery facilities are frequently faced with a mandate to accept increased wet weather flows, typically as a part of flow maximization to limit CSOs. In addition to the desktop design tools discussed previously, the maximum capacity of the existing primary treatment facilities (and the need for any additional facilities or facility modifications) could be evaluated through a full-scale demonstration test. Such tests provide the most direct way of evaluating capacity and needs, taking into account site-specific wastewater characteristics and clarifier design features. Because chemical addition could significantly improve clarifier performance, stress testing is typically done by subjecting an isolated test clarifier to various regimes of flows and chemical addition. Additional details on this are provided in Section 4.3.

4.2.2 Modification of Inlet–Outlet Structures and Use of Computational Fluid Dynamics Modeling

High-flow velocity at peak flow conditions will exacerbate uneven flow distribution caused by any hydraulic asymmetry in flow-splitting arrangements. Upflow distribution structures with flow velocities of no more than 0.3 m/s (1.0 ft/s) at peak flow are a preferred configuration (WEF, 2005).

In designing a primary clarifier, center wells (in circular clarifiers) and various inlet baffle arrangements (in rectangular clarifiers) are used to dissipate flow velocity (momentum) and prevent short-circuiting and sludge scouring. Computational fluid dynamics modeling tools are available (WEF, 2005) for the design or reconfiguration of primary clarifier internal structures for optimal performance at high flow conditions.

4.3 Chemically Enhanced Primary Treatment

The goal of CEPT is to deploy chemicals to improve particle settling in the primary clarifier. The rate of settling is governed by Stokes' Law, which states that the velocity of a settling particle is proportional to its diameter and density. The role of the chemicals is to increase particle diameter and density.

4.3.1 Role of Coagulants

The particles entering a treatment facility possess a small electrical charge. Respecting the laws of magnetism, it is difficult for these particles to coalesce and flocculate because of their natural tendency to repel each other. Consequently, the purpose of adding a coagulant is to neutralize the charge on the particles and render them suitable for flocculation. Examples of the most commonly used coagulants are metal salts, ferric chloride, alum, sodium aluminate, and polyaluminum chloride. Advances in polymer chemistry in the past decade have resulted in the development of cationic polymers that appear to work as well as metal salts in some instances.

4.3.2 Role of Polymer

Having created a body of neutralized particles, it remains for them to be flocculated such that the smaller and colloidal particles that were neutralized by coagulant addition form larger and denser particles. The agent of flocculation is typically an anionic polymer flocculant. Advances in polymer chemistry in the past decade have led to a wide range of anionic high-molecular-weight polymers being made available. A suitable polymer is added downstream of coagulant addition.

4.3.3 *Chemical Type and Dose Selection (Jar Testing)*

Although the influent BOD and TSS of raw wastewater fall into typical ranges at most treatment facilities, the chemistry of raw wastewater can differ significantly from facility to facility. A coagulant or polymer that works at one facility may not necessarily be successfully deployed at another facility. Therefore, it is important to conduct jar testing to verify the performance of specific chemicals for a specific application. The jar test can be considered as a small-scale simulation of what occurs at full scale. The results of jar tests are scalable to full-facility operating conditions. This simulation is also useful in determining the optimum dose of each chemical. A typical jar test assembly consists of six square sided jars operated in parallel with a common stirring mechanism as shown in Figure 12.8.

A test procedure can consist of the following: a large sample of raw wastewater is collected for each series of jar tests to allow comparison of different dosages and types of chemical for the same sample. The response to a test may be evaluated visually in some instances where rapid screening is desired, but measuring supernatant TSS concentration or turbidity will provide a definitive assessment, as shown in Figure 12.9. Initially, the approach might be to vary the dosage of a candidate polymer for a given dose of a coagulant. This can be repeated for different polymers. Once the best performing polymer had been identified, the response to varying the dose of the coagulant for a given dose of the selected polymer may be evaluated.

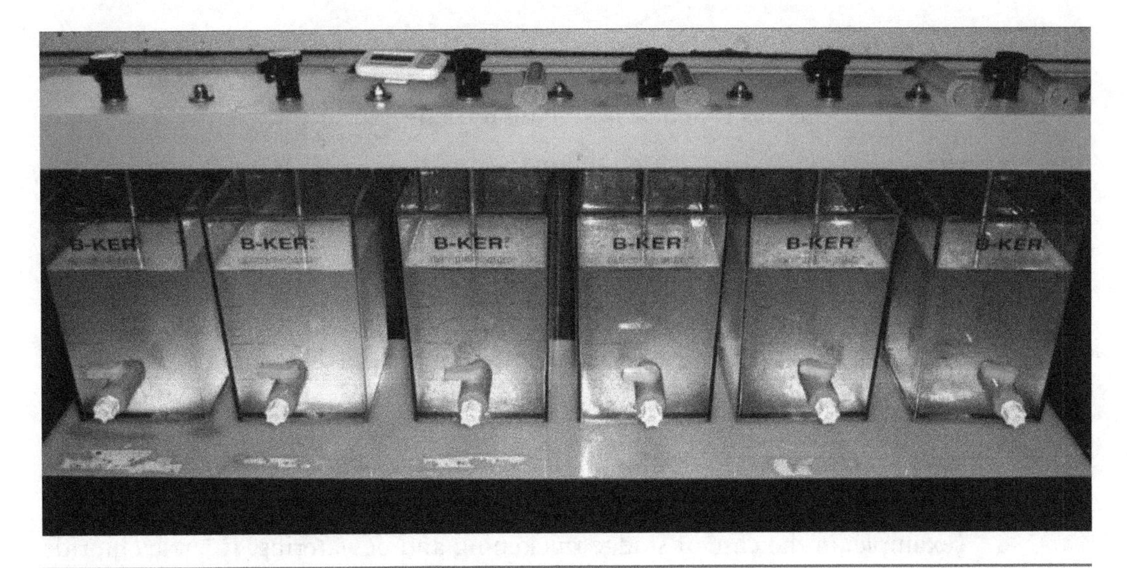

FIGURE 12.8 Six-jar test apparatus.

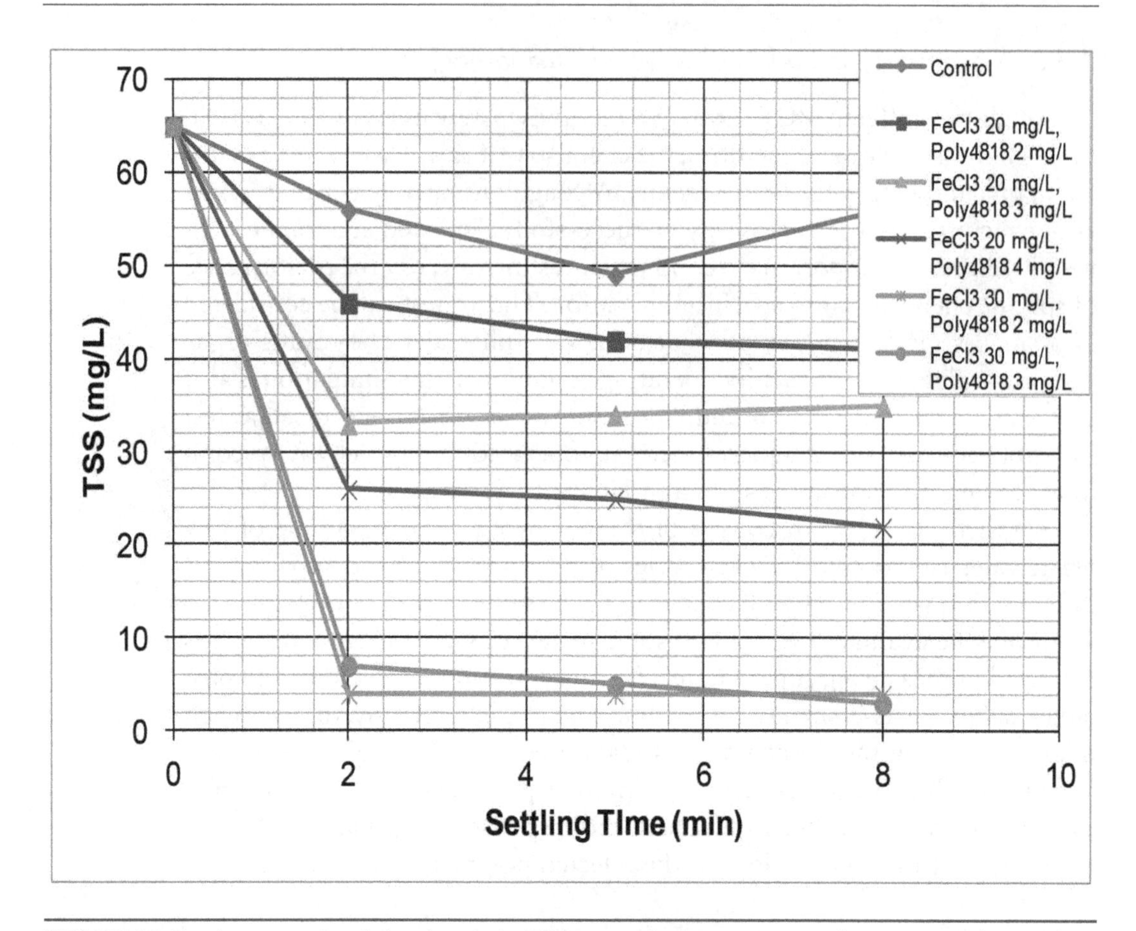

FIGURE 12.9 An example of the change in TSS in residual supernatant for a range of coagulant and polymer doses.

4.3.4 Chemical Dosing and Mixing Requirements in Chemically Enhanced Primary Treatment Systems

The equipment required for CEPT application is similar for pilot testing, full-scale demonstration, and application at full-scale operating facilities. The biggest difference between the different levels of scale is associated with the selection of pump sizes and pipe diameter. At the smaller scale, it can be difficult to find metering pumps that are robust enough to be used in the field while providing reliable flow measurement. Impurities in some coagulants can cause blockage in small-bore pipes. In a full-scale demonstration, a primary clarifier may be isolated and equipped to evaluate the effectiveness of CEPT (see the case study in Section 5.1). Chemical storage and delivery systems are typical of any chemical addition facility as, for example, in the case of sludge thickening and dewatering. If ferric chloride were to be used as the coagulant, delivery lines are typically double-lined and monitored for leaks.

Chemical dose, location, and the amount of mixing/turbulence at the point of addition have a large effect on the performance of the CEPT process. Coagulants should be delivered in an environment that is turbulent to promote rapid mixing and dispersion of the coagulant. Examples of such installations include pump intakes, aerated grit removal tanks, an ogee hydraulic jump, at the entry to a Parshall flume, and the use of an induction mixer or an air sparging device. The flocculant, too, needs to be added at a location that is turbulent to provide rapid dispersion, but it is important that this turbulence dissipates quickly and that the newly formed floc does not experience significant hydraulic disturbances downstream of the polymer addition point to avoid destruction of the floc. Figure 12.10 shows the polymer addition assembly at the Columbia Boulevard Treatment Plant in Portland, Oregon. It is located just at the point where the primary influent emerges from the main influent pipe into the distribution channel of the wet weather primary clarifiers. Similar types of locations should be sought for the polymer addition point.

4.3.5 Instrumentation and Process Control

Instrumentation and controllers for chemical delivery systems are similar to those deployed in sludge thickening and dewatering applications. The rate of addition of coagulant and flocculant is typically flow-paced to maintain the desired dose. Control of the overall CEPT process is currently evolving; the best response characteristic for monitoring primary clarifier effluent quality

FIGURE 12.10 Polymer addition point at the Columbia Boulevard facility.

is the effluent turbidity. Online turbidity sensors are reliable and may be used to collect continuous turbidity data in the influent and effluent.

5.0 CASE STUDIES

5.1 King County's South and Brightwater Plants, Seattle, Washington

King County's 144-ML/d (38-mgd) Brightwater Treatment Plant treats wet weather flows with CEPT. Dry weather flow is treated with MBR technology (Melcer et al., 2004). To minimize the cost of membranes, peak wet weather flows are directed to CEPT and then combined with MBR effluent before disinfection. Full-scale demonstration of CEPT was conducted at King County's South Plant during the winter of 2004–2005 to collect data for the design of the Brightwater CEPT system. High removals of TSS (80 to 90%) and BOD (58 to 68%) (Figure 12.11) were achieved with sequential

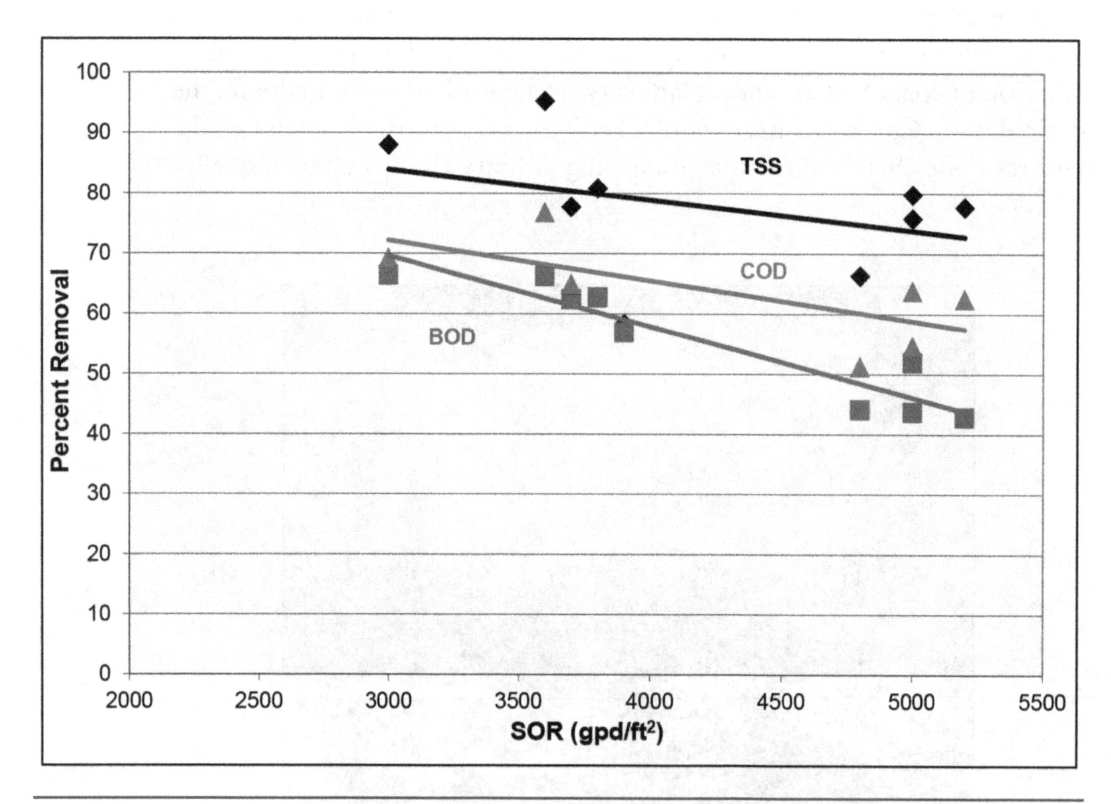

FIGURE 12.11 Performance of South Plant CEPT primary clarifiers with 50 to 60 mg/L $FeCl_3$, 10 to 15 mg/L polyaluminum chloride, and 0.3 to 0.5 mg/L anionic polymer (Melcer et al., 2005).

dual-coagulant (ferric chloride, polyaluminum chloride) and anionic polymer addition at a primary clarifier peak hour SOR of 147 m/h (3600 gpd/sq ft) (Melcer et al., 2005). Without chemicals, TSS and BOD removals at peak SOR were approximately 50 and 25%, respectively. At higher SORs of 204 m/h (5000 gpd/sq ft), TSS and BOD were approximately 65 and 40%, respectively, in the CEPT clarifiers.

Biochemical oxygen demand removal was shown to depend on the particulate BOD fraction. For the South Plant, approximately 62% of the influent wastewater BOD is particulate. Without chemicals, BOD removals were less than 62%. However, up to 68% BOD removal was achieved at SORs of less than 163 m/h (4000 gpd/sq ft), indicating that all particulate BOD, and a small portion of colloidal BOD, had been removed. Lower BOD removals at SORs greater than 163 m/h (4000 gpd/sq ft) were attributable to the inability of the primary clarifiers to capture and settle particulate BOD at the higher flows.

The Brightwater CEPT installation was brought online during 2012 and performance has been identical to the South Plant demonstration data.

5.2 Columbia Boulevard Wastewater Treatment Plant, Portland, Oregon

The capacity of the secondary treatment system at the city of Portland's Columbia Boulevard Wastewater Treatment Plant is limited to 380 to 455 ML/d (100 to 120 mgd). At the time that dry weather flow primary clarifiers were installed, flows in excess of this were directed to 50-year-old primary clarifiers that are now retained only for processing wet weather flows and are referred to as *wet weather primary clarifiers* (WWPCs). The city has installed two large CSO interceptors increasing the peak flow to the Columbia Boulevard facility from 1325 to 1700 ML/d (350 to 450 mgd) and elevating the WWPC peak SOR to 122 m/d (3000 gpd/sq ft). In 2008, new regulatory TSS and BOD removals of 70 and 50% removal, respectively, were anticipated for the WWPCs. Historical BOD and TSS removals by the WWPCs were relatively low even at low SORs; Figure 12.12 shows that, in 2006, the proposed removals could be achieved at a relatively low 24.5 m/d (600 gpd/sq ft) for TSS and 49 m/h (1200 gpd/sq ft) for BOD and were unlikely to meet the required removals at the higher SOR condition (Melcer et al., 2010).

Chemically enhanced primary treatment was investigated to take advantage of existing primary clarifiers that can be made to operate at higher efficiencies with chemical addition during high-flow scenarios, precluding the need to invest in high-rate clarification or additional conventional primary clarifiers. Bench-scale jar tests were conducted in 2008 to determine

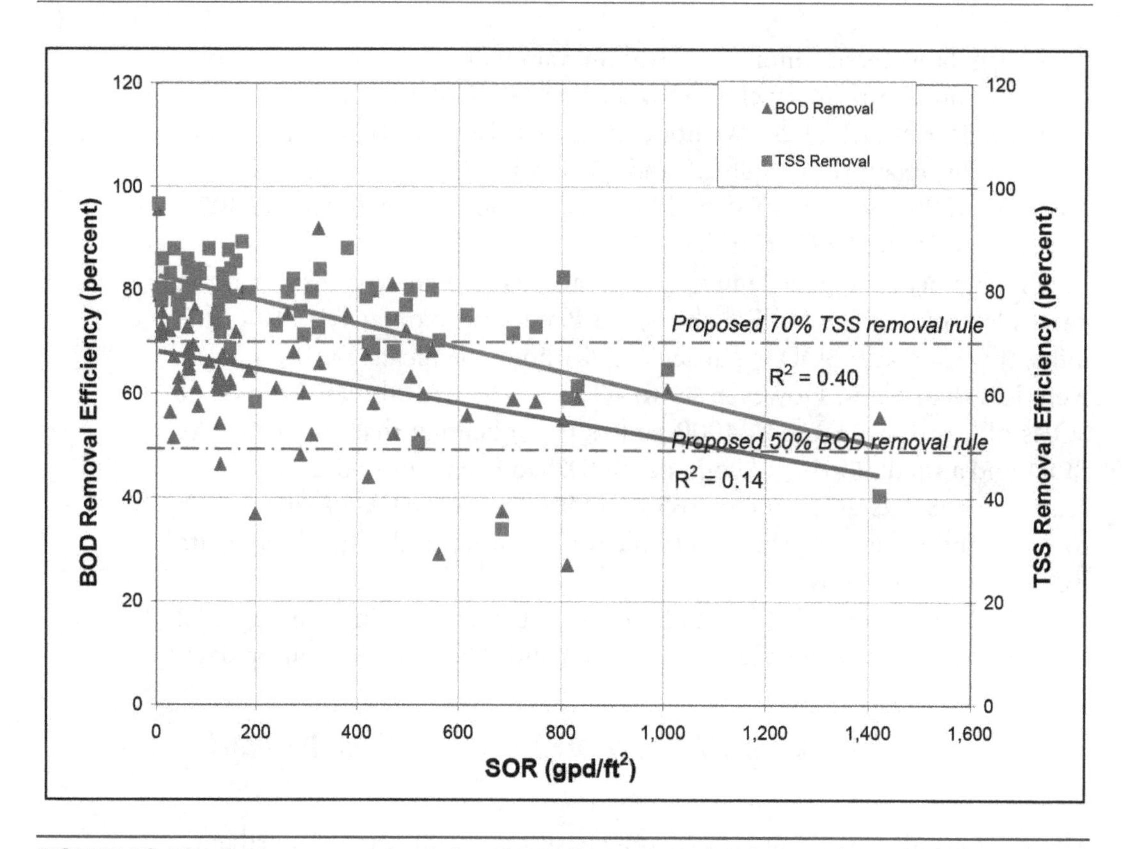

FIGURE 12.12 Performance of existing WWPCs at the Columbia Boulevard facility in 2006 (Melcer et al., 2010).

which chemicals were suitable for the WWPC influent and at what dosages. Ferric chloride and anionic polymer were identified as the most appropriate chemicals (Melcer et al., 2010). In 2009, they were tested in the field during wet weather events to verify the bench-scale test results and to evaluate the best hydraulic location for introducing them to the primary influent. The maximum SOR experienced was 118 m/d (2900 gpd/sq ft); unfortunately, significant high-flow events were not experienced during all four events observed. With the exception of the March 15, 2009, event, the CEPT system consistently achieved greater than 80% TSS removal and greater than 65% BOD removal during the trial (greater than the target levels). Removal performance appeared to be related to the degree of influent dilution. During the trial, the ferric chloride dose was gradually reduced from 50 to 25 mg/L. The polymer dose was more stable, with the best results observed at a concentration of 1.2 to 1.3 mg/L.

The full-scale system was installed in 2012, but has not yet experienced significant wet weather events because of the drought conditions in the Portland area during the winter of 2012 to 2013.

Similar investigations have been conducted at the Metropolitan District C's Hartford, Connecticut, water resource recovery facility (Newman et al., 2013) and are underway at the Northeast Ohio Regional Sewer District's (NEORSD) three Cleveland, Ohio, water resource recovery facilities (Melcer et al., 2012). Modified CEPT facilities are currently in design in Hartford and under construction at the NEORSD facilities.

6.0 REFERENCES

Barnard, J. L. (1984) Activated Primary Tanks for Phosphate Removal. *Water SA*, **10** (3), 121–126.

Crow, J.; Smyth, J.; Bucher, B.; Sukapanotharam, P. (2012) Treating CSOs Using Chemically Enhanced Primary Treatment With and Without Lamella Plates—How Well Does It Work? *Proceedings of the 85th Annual Water Environment Federation Technical Exhibition and Conference* [CD-ROM]; New Orleans, Louisiana, Sept 29–Oct 3; Water Environment Federation: Alexandria, Virginia.

Fitzpatrick, J. D.; Bradley, P. J.; Duchene, C. R.; Gellner, J., II; O'Bryan, C. R., Jr. Ott, D.; Sandino, J.; Tabor, C. W.; Tarallo, S. (2012) Preparing for a Rainy Day—Overview of Treatment Technology Options for a Wet-Weather Flow Management. *Proceedings of the 85th Annual Water Environment Federation Technical Exhibition and Conference* [CD-ROM]; New Orleans, Louisiana, Sept 29–Oct 3; Water Environment Federation: Alexandria, Virginia.

Fitzpatrick, J. D. Long, M.; Wagner, D.; Middlegrough, C. (2008) Meeting Secondary Effluent Standards at Peaking Factors of Five or Higher. *Proceedings of the 81st Annual Water Environment Federation Technical Exhibition and Conference* [CD-ROM]; Chicago, Illinois, Oct.18–22; Water Environment Federation: Alexandria, Virginia.

Krugel, S.; Melcer, H.; Hummel, S., Butler, R. (2005) High Rate Chemically Enhanced Primary Treatment as a Tool for Wet Weather Plant Optimization and Re-Rating. *Proceedings of the 78th Annual Water Environment Federation Technical Exhibition and Conference* [CD-ROM]; Washington, D.C., Oct 29–Nov 2; Water Environment Federation: Alexandria, Virginia.

Leffler, M. R.; Harrington, J. (2001) SSO Elimination through Expanded Primary Treatment Capacity and Blended Effluent. *Proceedings of the 74th Annual Water Environment Federation Technical Exposition and Conference* [CD-ROM]; Atlanta, Georgia, Oct 13–17; Water Environment Federation: Alexandria, Virginia.

Melcer, H.; Davis, D.; Xiao, S.; Shaposka, H.; Ifft, J.; Bucurel, N.; Land, G. (2012) Wet Weather Flow Treatment with a Difference: Novel Ideas for Applying Chemically Enhanced Primary Treatment with High Rate Disinfection. *Proceedings of the 85th Annual Water Environment Federation Technical Exhibition and Conference* [CD-ROM]; New Orleans, Louisiana, Oct 2–6; Water Environment Federation: Alexandria, Virginia.

Melcer, H.; Ciolli, M.; Lilienthal, R.; Ott, G.; Land, G.; Dawson, D.; Klein, A.; Wightman, D. (2010) Bringing CEPT Technology into the 21st Century. *Proceedings of the 83rd Annual Water Environment Federation Technical Exhibition and Conference* [CD-ROM]; New Orleans, Louisiana, Oct 2–6; Water Environment Federation: Alexandria, Virginia.

Melcer, H.; Krugel, S.; Butler, R.; Carter, P.; Land, G. (2005) Alternative Operational Strategies to Control Pollutants in Peak Wet Weather Flows. *Proceedings of the 78th Annual Water Environment Federation Technical Exhibition and Conference* [CD-ROM]; Washington, D.C., Oct 29–Nov 2; Water Environment Federation: Alexandria, Virginia.

Melcer, H.; Tam, P.; Tse, R.; Burke, P.; Hummel, S. (2004) New Technology for a New Plant: Brightwater Membrane Bioreactor Process. *Proceedings of the 77th Annual Water Environment Federation Technical Exhibition and Conference* [CD-ROM]; New Orleans, Louisiana, Oct 2–6; Water Environment Federation: Alexandria, Virginia.

Newman, D.; Melcer, H.; Davis, D.; Pepe, L.; Winn, R.; Nascimento, D.; Tyler, T. (2013) At the Nexus of Process and Design: Optimizing a Wet Weather Treatment System. *Proceedings of the 86th Annual Water Environment Federation Technical Exhibition and Conference* [CD-ROM]; Chicago, Illinois, Oct 7–9; Water Environment Federation: Alexandria, Virginia.

U.S. Environmental Protection Agency (1978) *Field Manual for Performance, Evaluation and Troubleshooting at Municipal Wastewater Treatment Facilities;* EPA-430/9-78-001; U.S. Environmental Protection Agency: Washington, D.C.

U.S. Environmental Protection Agency (1999) *Combined Sewer Overflow Technology* Fact Sheet: Retention Basins; EPA-832/F-99-042; U.S. Environmental Protection Agency: Washington, D.C.

U.S. Environmental Protection Agency (2000) *Retrofitting Control Facilities for Wet-Weather Flow Treatment;* Research Report, EPA-600/R-00-020; U.S. Environmental Protection Agency: Washington, D.C.

U.S. Environmental Protection Agency (2003) *Wastewater Technology Fact Sheet: Screening and Grit Removal;* EPA-832/F/03-011; U.S. Environmental Protection Agency: Washington, D.C.

Wahlberg, E. J.; Wang, J. K.; Merill, M. S.; Morris, J. L.; Kido, W. H.; Swanson, R. S.; Finger, D.; Philips, D. A. (1997) Primary Sedimentation: It's Performing Better Than You Think. *Proceedings of the 70th Annual Water Environment Federation Technical Exposition and Conference;* Chicago, Illinois, Oct 18–22; Water Environment Federation: Alexandria, Virginia; pp. **1**, 731.

Water Environment Federation (2005) *Clarifier Design*, 2nd ed.; WEF Manual of Practice No. FD-8; Water Environment Federation: Alexandria, Virginia.

Water Environment Federation; American Society of Civil Engineers; Environmental & Water Resources Institute (2010) *Design of Municipal Wastewater Treatment Plants*, 5th ed.; WEF Manual of Practice No. 8; ASCE Manuals and Reports on Engineering Practice No. 76; Water Environment Federation: Alexandria, Virginia.

13

Biological Treatment

W. James Gellner, P.E.

1.0 INTRODUCTION

The biological treatment system at a water resource recovery facility (WRRF) is the backbone of the facility's effectiveness in meeting permit limits. Effective management of the biological treatment system during wet weather events is often the single most important factor in reliably meeting weekly or monthly permit limits. If biomass and sensitive nutrient removal populations are washed out of the mainstream process resulting from upset during wet weather, facility performance can be adversely affected for an extended period. Therefore, the biological system is an important consideration in most facility improvement projects involving wet weather capacity. For facilities in which increased treatment capacity is needed to comply with regulatory requirements, modifications to the biological treatment system are often cost-effective. This is especially true when ample settling/sedimentation infrastructure is already available.

This chapter focuses on available wet weather flow and load management strategies and provides the reader with a guidance on considerations for each as well as design considerations for increasing capacity of the biological treatment system. This chapter is not intended to be a comprehensive design guide for biological treatment systems. Detailed design guidance for biological systems is provided in *Design of Municipal Wastewater Treatment Plants* (WEF et al., 2010).

2.0 EFFECTS OF WET WEATHER ON BIOLOGICAL TREATMENT

Wet weather events can have a significant effect on the biological treatment system. An overview is provided in this section.

2.1 Changing Influent Characteristics

2.1.1 Flow

Increases in facility flow are common for both separate and combined collection systems during wet weather. The magnitude and duration of the flow increase is highly dependent on the size, type, and characteristics of the collection system tributary to the facility. Typically, combined systems have higher peaks relative to average dry weather flows than separate systems. Peaks with flows greater than 10 times those observed during dry weather are not uncommon for combined systems. Common peaking factors in separate systems are 3 to 5 times the average the dry weather flow. The actual peak flow depends heavily on a host of factors, including the size of the collection system, age, and proximity to groundwater and other surface water.

The timing of peak flows relative to the start of rain events in the area of the collection system is also an important consideration. The rate of change of flow can also be different depending on rainfall location and intensity and the size and characteristics of a collection system. The lag time between the wet weather event and its manifestation because of higher flows to the WRRF influences the required response time in making adjustments in operation. More often than not, operational adjustments must be made well in advance of rain events to effectively transition to wet weather operations. Snowmelt events can also increase groundwater levels, stream flows, and infiltration/inflow to collection systems.

2.1.2 Temperature

Wet weather flows are often accompanied by a drop in wastewater temperature. In some instances, the drop in wastewater temperature can be significant relative to ambient conditions, particularly in areas where snowmelt affects wet weather flows (Katehis et al., 2011). When drops in wastewater temperature occur, process kinetics are reduced, sometimes drastically. When this reduction in process kinetics is coupled with reduced hydraulic retention time, treatment efficiency across the biological process can be significantly compromised.

2.1.3 Loads

Similar to the discussion on peak flows, increases in loadings from wet weather events vary from system to system. Additional loads are typically associated with the transport of surface nutrients and scoured solids from erosion and other sources. In instances where surface stream elevations exceed collection system manholes and pipe inverts, river sediment can often also be carried into the system during a wet weather event.

As flows increase at the start of a wet weather event, the flushing of deposited solids in the sewer system will often increase all loads, including organics and nutrients. Most often, this increase results from short-term scouring of solids from the collection system when flow velocities increase. Depending on the characteristics of the collection system, soluble organic loading can also increase during this period. The magnitude and characteristics of the increase are unique at each facility and depend on sewer system characteristics, storm intensity, time from the last event, and other factors.

This increase is often termed *first flush* and can have significant effects on a WRRF, particularly in preliminary treatment. Grit and screening systems can be overwhelmed by large solids loads if those facilities have not been designed to accommodate large and rapid increases in loadings. An example of changes in influent solids during the initial hours of a wet weather event are shown in Figure 13.1. Depending on the performance of preliminary and primary treatment, a portion of these increases in loadings can reach the biological system.

2.1.4 Biochemical Oxygen Demand/Total Suspended Solids

Suspended solids often increase dramatically during first flush. The composition of biochemical oxygen demand (BOD) and chemical oxygen demand (COD) may also change because of changes in collection system temperature and retention time. For example, particulate fractions of BOD and COD may increase. If there are areas of the collection system where significant deposition of organics occur during dry weather periods, soluble constituents will also increase during the first flush. After the initial spike of concentration (and loading), the concentrations of most constituents decrease sharply (see Figure 13.1). At facilities that are influenced by surface water intrusion to the collection system, a higher fraction of inert suspended solids

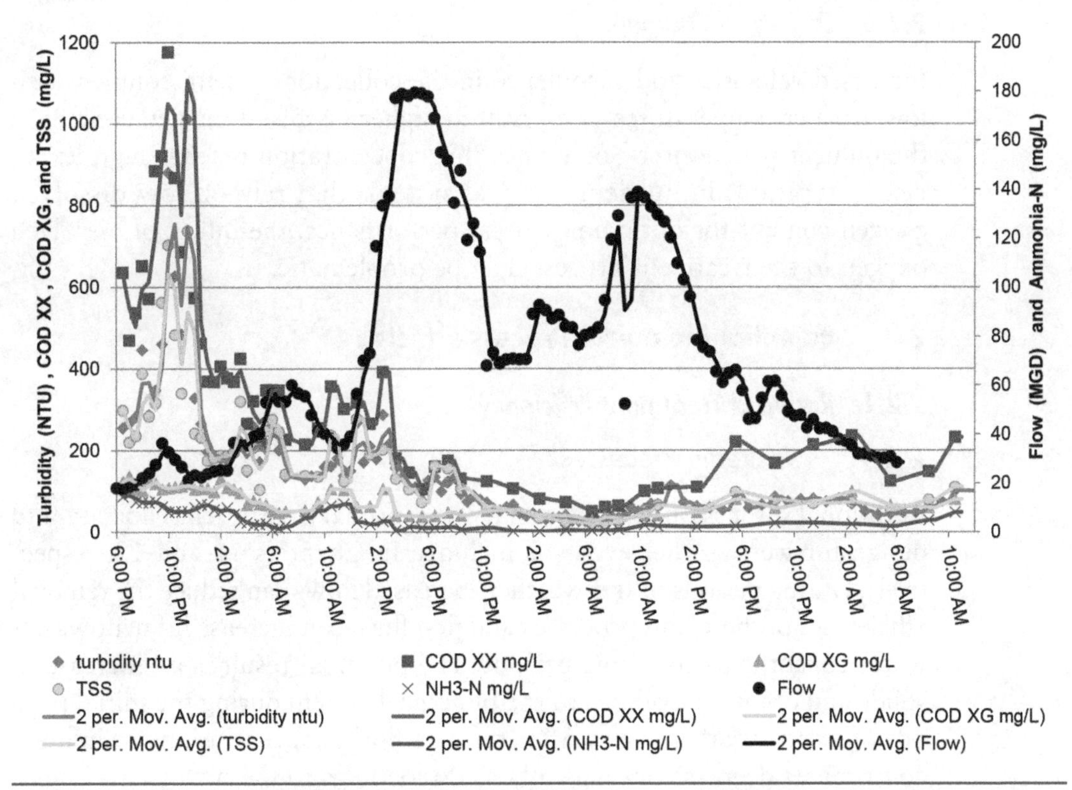

FIGURE 13.1 Example of first flush concentration at a WRRF (Little Miami Wastewater Treatment Plant, Metropolitan Sewer District of Greater Cincinnati, October 2011).

may enter the facility during the wet weather event. When these inerts pass through preliminary and primary treatment, they can increase the inert solids concentration in the biological system. When that system is operated to maintain a set mixed liquor suspended solids (MLSS), the amount of active biomass in the system can be reduced at equivalent MLSS concentrations, effectively lowering treatment capacity.

2.1.5 Water Chemistry

The dilution of incoming wastewater flows with rainwater changes influent water chemistry. Often, alkalinity is decreased because of the dilution of raw wastewater. Changes in influent chemistry can affect solids settleability, pH of treatment processes, and biological kinetics in systems with typically low background alkalinity. In tidal areas and areas where snow is prevalent, wet weather events can also result in significant increases in salinity because of road salt and coastal water intrusion (Katehis et al., 2011). Salinity increase may also affect the flocculation properties of activated sludge.

2.1.6 Dissolved Oxygen

Increased velocities and turbulence in the collection system, coupled with lower water temperatures, can result in higher dissolved oxygen content in the influent wastewater. Sometimes this concentration reaches high levels, near saturation. In nutrient removal systems that rely on low dissolved oxygen content for optimum process performance, the influx of dissolved oxygen to the treatment process may be problematic.

2.2 Potential Biological Process Effects

2.2.1 Reduced Treatment Efficiency

2.2.1.1 Upstream Processes

Additional discussion on preliminary and primary treatment capacity and design for wet weather events is included in Chapters 11 and 12, respectively. During periods of wet weather, increased flows can reduce the removal efficiencies of these unit processes and first flush can increase influent wastewater constituents for short periods. The potential result is an increase in solids and organics loadings to the biological system during the initial periods of a wet weather event. After the first flush, concentrations of influent constituents decrease dramatically as do loadings to the biological system.

2.2.1.2 Chemical Oxygen Demand/Biochemical Oxygen Demand

Short detention times and relatively low sludge ages are required in a biological treatment system for COD/BOD removal. Decreases in COD and BOD removal efficiency can occur in wet weather when detention times are short (i.e., on the order of minutes). Very short detention times can be used for wet weather biological treatment, particularly when step feed/contact stabilization is used. During short detention time treatment, the absorption of particulate and colloidal organics into the activated sludge floc can be decreased, although typically not to levels that affect compliance with the permit. However, the designer should be aware of this potential influence.

2.2.1.3 Ammonia Removal

Permit requirements for nitrification are common for WRRFs. The kinetics of nitrification in biological treatment processes are well understood and are influenced by many factors, including dissolved oxygen, temperature, detention time, active nitrifier biomass in the system, and pH. Nitrification is most commonly affected during wet weather events through reduced detention time in biological reactors. In systems where wet weather strategies are used to retain biomass in the biological system, the mixed liquor biomass inventory in the biological reactor is sometimes reduced, decreasing the amount of biomass in contact with the wastewater flow and decreasing removal efficiencies. The decrease in removal efficiency is often tempered by reductions in ammonia concentrations after the first flush. However, the amount of decrease in nitrification is an important consideration during facility design, particularly when the facility has a stringent mass-based discharge limit.

2.2.1.4 Nutrient Removal

Discrete biological treatment zones with unique operating conditions are necessary for effective biological nutrient removal. For example, phosphorus removal depends on the creation of an anaerobic environment, with the anaerobic zones free of both nitrate and dissolved oxygen to maintain phosphorus accumulating organism (PAO) populations and to drive biological phosphorus removal. Similarly, nitrogen removal relies on anoxic environments with little dissolved oxygen. Both processes are also reliant on the amount of active biomass in the system and hydraulic retention time in those dedicated zones for effective treatment.

Wet weather events create conditions that reduce the efficiency of nutrient removal systems. Increased influent dissolved oxygen, increased flowrates, and reduced influent temperature decrease performance in nutrient removal systems. Reduced active biomass is also a potential influence in facilities using a biological treatment wet weather operating strategy. Similar to consideration of nitrification, the designer must consider the effects of reduced nutrient removal performance on mass-based discharge permit requirements. When mass loadings of nutrients are increased during first flush wet weather periods, the ability of the facility to quickly return to typical dry weather treatment levels and to operate in an optimized fashion during periods of dry weather are often the key to maintaining discharge permit compliance.

Changes to COD composition as a result of wet weather events may also affect performance of nutrient removal processes that are dependent on the readily biodegradable fraction. These processes include biological phosphorus removal and denitrification.

2.2.1.5 Increased Biosolids Quantities and Yield

Observed yield and biosolids quantities can also increase during wet weather. If wet weather flows and associated loads are sustained for long periods, wasting must increase to maintain target mixed liquor concentrations in the biological reactor. Maintaining MLSS targets during a wet weather event may require increased amounts of wasting to the solids handling system.

During wet weather, this can temporarily decrease the effective sludge age of the biological system. Additional solids can also be generated when physical/chemical treatment is used to augment biological treatment. Solids handling systems should be designed to handle short-term increases in sludge quantities resulting from wet weather events.

2.2.1.6 Secondary Clarifier Efficiency

Increased flow and increased solids loading rates to secondary clarifiers will reduce removal efficiency resulting from higher upflow velocities and reduced detention time. Increases in sludge blankets reduce the effective clarification volume of the clarifier. If not managed properly, washout of biomass can occur when sludge blankets are allowed to increase to significant depths approaching the effluent weirs.

2.2.2 Changes in Solids Inventory and Location

Wet weather events change the characteristics of the biological solids inventory and its location. Where wet weather strategies are not used, undesirable changes may occur as a result of the increased flow and solids loading to the biological system. In facilities using wet weather strategies, changes occur intentionally as a result of the strategy used to mitigate undesirable inventory changes.

When increases in facility flow are sent through the biological system, the solids loading to the downstream solids separation system is increased. If return activated sludge (RAS) rates are not increased, solids typically begin to accumulate in the clarifiers and more biomass is shifted to the clarifiers, reducing mixed liquor at the upstream end of the biological reactor. Because no aeration is being provided in the clarifiers, these solids are not productive relative to treatment. In facilities using wet weather strategies, the locations of "accumulated solids" are deliberately manipulated to optimize system performance during rainfall events.

2.3 Performance Objectives for Wet Weather Biological Treatment

2.3.1 Discharge Permit Compliance

During wet weather events, reductions in treatment efficiency can negatively affect discharge permit compliance. Most WRRFs are required to meet mass-based monthly and weekly effluent requirements. In addition, they must also comply with the secondary treatment standard of 85% monthly average removal of BOD and total suspended solids (TSS).

When increases in pollutant mass discharge occur during wet weather events, the facility must quickly return to stable operation during subsequent dry weather and operate effectively to meet permit conditions. The facility must operate at higher treatment efficiencies during these dry weather periods to compensate for pollutant mass discharges during wet weather. The potential effects of wet weather events on mass discharge requirements and the capacity to operate at more efficient treatment levels during dry weather must be considered during design.

As a simple example, consider a facility with a 18.9-ML/d (5-mgd) rated capacity and a monthly average ammonia discharge limit of 1 mg/L. This ammonia discharge limit corresponds to a total mass loading during a 30-day period of

$$18.9 \text{ ML/d} \times 1 \text{ mg/L} \times 30 \text{ days} = 567 \text{ kg } NH_3\text{-N/mo} \ (1250 \text{ lb } NH_3\text{-N/mo})$$

If a wet weather event occurs that averages 37.8 ML/d (10 mgd) for a period of 3 days and ammonia concentrations in the discharge reach 3 mg/L, the total mass discharge during that wet weather event is

$$37.8 \text{ ML/d} \times 3 \text{ mg/L} \times 3 \text{ days} = 340 \text{ kg } NH_3\text{-N discharged during event } (750 \text{ lb } NH_3\text{-N})$$

Based on mass loading alone, the facility would need to operate well under the permit concentration level for the remaining 27 days of the permit. Following the aforementioned example, the facility would need to achieve an average concentration of 0.44 mg/L over the remaining days in that monitoring month if the facility operates near capacity, as follows:

$$(567 \text{ kg } NH_3\text{-N} - 340 \text{ kg } NH_3\text{-N}) / 27\text{d} / 18.9 \text{ ML/d} = 0.44 \text{ mg/L average concentration}$$

Many facilities also have a weekly loading limit for effluent discharges. Continuing the example of the 18.9-ML/d facility, ammonia concentration limits for maximum weekly discharge may be as high as 1.5 times the monthly concentration limit on a mass basis. So, for a 7-day period, the average effluent concentration may be 1.5 mg/L. For the week, the total allowable mass discharge would be

$$18.9 \text{ ML/d} \times 1.5 \text{ mg/L} \times 7 \text{ days} = 198 \text{ kg } NH_3\text{-N per week} \ (438 \text{ lb } NH_3\text{-N per week})$$

Using the preceding hypothetical wet weather event, the facility would have discharged more than the weekly permit allowance during the wet weather event and would have to record a weekly ammonia violation for that month.

The aforementioned example illustrates the potential effect of wet weather on a facility's ability to meet a permit. The designer must consider the effects of reduced treatment efficiency as part of the evaluation of wet weather strategies to ensure that the wet weather strategy used meets the effluent compliance needs of the facility under different operating conditions.

2.3.2 Prevent Biomass Washout

Reduced secondary clarifier efficiency resulting from higher hydraulic and solids loading is expected during wet weather events. If solids are released in large amounts from the clarifier overflow, this can lead to complete washout of active biomass. If the nitrifier population is washed out of the system, there is a long recovery period, typically 3 times the solids retention time or more to return to stable nitrification. Obviously, the effect of this type of solids loss is significant and could lead to noncompliance, especially for nitrification. The younger biomass during the regrowth period may also be harder to settle, leading to solids separation problems during operation in dry weather. The facility would also be more susceptible to facility upset during a subsequent wet weather event that occurs during the biomass regrowth period.

An effective wet weather strategy provides the WRRF with tools to prevent biomass washout. The facility must couple these tools with an effective operations strategy to fully protect against large losses of biomass.

2.3.3 Maintain Biological Population

Facilities with stringent nutrient removal requirements must maintain stable operation of the biological nutrient removal process during dry weather conditions and prevent sensitive population washout during a wet weather event. This is particularly important for biological phosphorus removal, resulting from the need to nurture and maintain a stable population of PAOs in the biological system. Similar to the washout example given previously for facilities with ammonia limits, the washout of PAOs during a wet weather event, either because of loss of solids or resulting from the inability to maintain cyclic anaerobic/aerobic conditions for the population, can mean the need for chemical feed for phosphorus precipitation and difficulty in re-establishing an effective PAO population.

3.0 OPTIONS FOR WET WEATHER BIOLOGICAL TREATMENT

3.1 Introduction

This section contains a discussion of treatment process strategies that can be used to increase the treatment capacity of the secondary treatment process during wet weather flow events. Basic configurations of the process options are discussed herein. The reader/designer should consider site-specific limitations when applying these generic concepts for specific projects.

3.2 Step Feed/Contact Stabilization

Contact stabilization and *step feed* are terms used to describe the wet weather strategy of moving the feed of a portion or all of the incoming flow to a downstream cell in a plug flow reactor. *Step feed* is the term used most often to describe wet weather strategies in plug flow tanks. However, use of step feed converts single-pass tanks to a contact stabilization process for a short period of time during wet weather. For the discussion included herein, a distinction will be made on the use of these terms for wet weather. The dry weather description of these process variations is provided first, followed by a description of their use for wet weather.

The contact stabilization process was originally developed and used as a dry weather process for BOD removal with a compact footprint. A typical dry weather contact stabilization schematic is shown in Figure 13.2.

The contact stabilization process consists of a short detention time contact tank, followed by clarification. Contact times within the contact tank are small, typically 30 to 60 minutes and up to 2 hours maximum. Return sludge is returned through a reaeration or stabilization tank. With more concentrated flows, the reaeration tank provides a larger hydraulic retention time, ranging from 3 to 8 hours. Rapid biosorption of particulate, colloidal, and soluble organics occurs in the contact tank. Those organics are further stabilized in the longer detention time reaeration tank (Wang et al., 2009).

Step feed is typically used for dry weather applications for the following two purposes:

1. Spreading of incoming organic load to equalize oxygen demand along the length of a plug flow aeration tank and
2. Adding influent wastewater at multiple anoxic zones along the basin to improve denitrification.

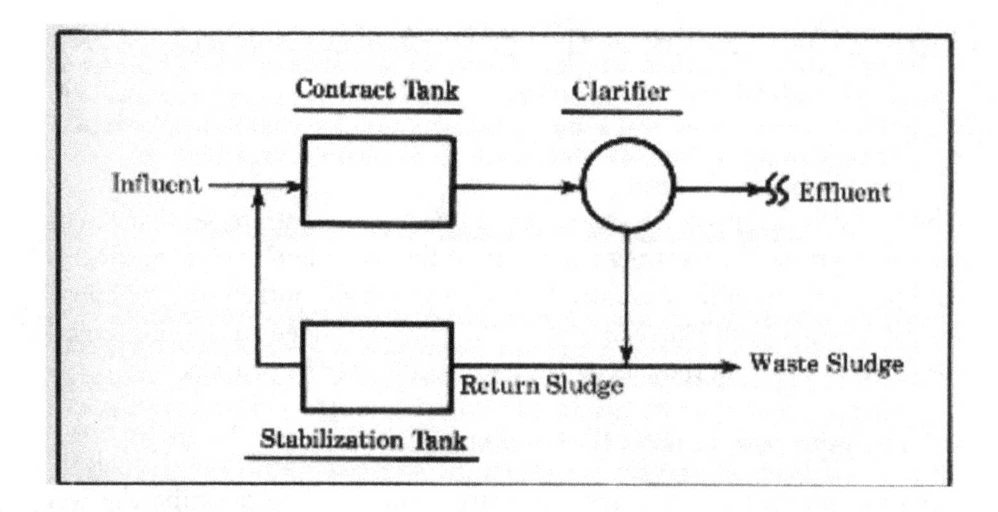

FIGURE 13.2 Schematic of a dry weather contact stabilization process.

An example of a dry weather step-feed process configuration previously used at the Newtown Creek Wastewater Treatment Plant (New York) for BOD removal is shown in Figure 13.3 (Fortin et al., 2007). Dry weather step feed for enhancing denitrification used at the Wards Island Wastewater Treatment Plant, Battery E (New York), is shown in Figure 13.4 (Gellner et al., 2012).

When step feed is used for wet weather, it is typically characterized by feeding a larger amount of the influent wastewater to downstream portions of a plug flow aeration tank (or subsequent passes of a multipass tank) or by feeding a greater percentage of influent wastewater to multiple-pass basins where step feed is used for dry weather. Return activated sludge feed is continued in the upstream portions of the aeration tank, most often to the locations originally used for dry weather. Examples of step feed in plug flow reactors and multipass tanks are shown in Figures 13.5 and 13.6, respectively (Pitt et al., 2007).

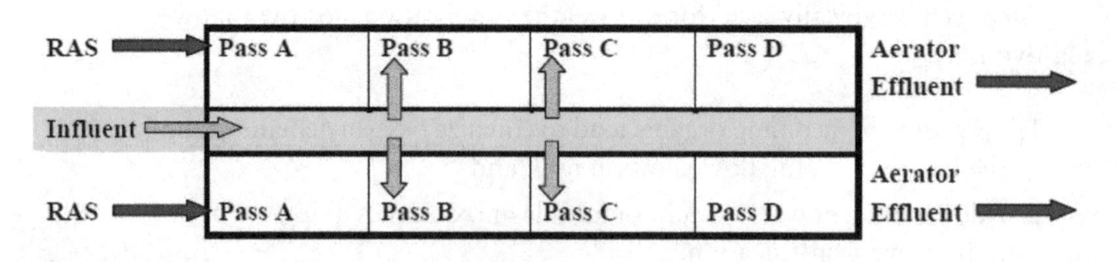

FIGURE 13.3 Dry weather step-feed configuration used at the Newtown Creek facility (Fortin, et al., 2007).

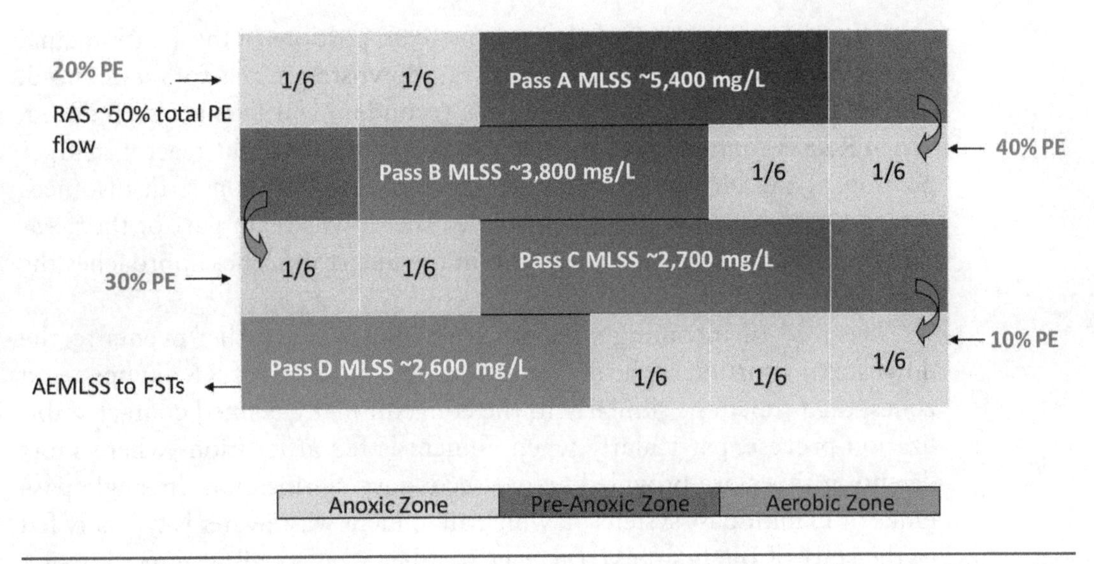

FIGURE 13.4 Dry weather step-feed configuration for denitrification used at the Bowery Bay Wastewater Treatment Plant (Gellner et al., 2012).

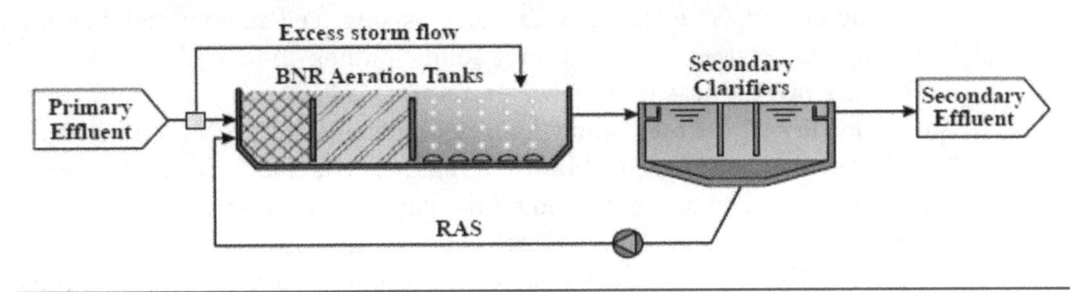

FIGURE 13.5 Example of step feed in a single-pass plug flow reactor (Pitt et al., 2007).

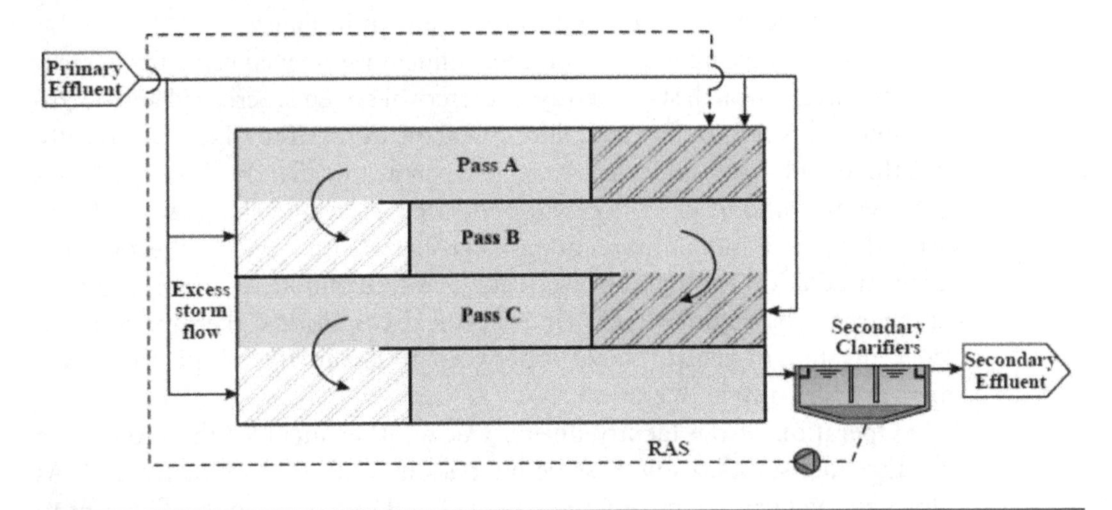

FIGURE 13.6 Example of step feed in multiple-pass plug flow reactors (Pitt et al., 2007).

When incoming flow is fed to downstream portions of the aeration tank, the resulting MLSS concentration in the downstream reactors is reduced. The resulting solids loading rate to the secondary clarifiers is also reduced. When RAS is continually fed to the upstream portions of the reactor, a larger portion of the biomass accumulates in the upstream zones. In instances where all of the incoming flow is fed to the downstream part of the aeration tank, the biomass concentration in the upstream zones approaches the RAS concentration over time.

The feed of incoming wastewater during wet weather events to the downstream portion of the reactors, with accumulation of RAS in upstream zones, then functions similarly to the conventionally defined contact stabilization process, particularly when influent is fed at locations where short detention times are provided before secondary clarification. In single-pass tanks or in multipass systems in which all influent wastewater is typically fed to the start of the first pass, this wet weather strategy also mimics contact stabilization. Where multiple feed locations are used during dry weather, changes in feed for wet weather operation more closely resemble the step feed process.

The city of Greencastle, Indiana, uses step feed/contact stabilization during wet weather events to limit solids loading to secondary clarifiers. The City of Greencastle Wastewater Treatment Plant is a secondary treatment facility with a rated annual average capacity of 10.6 ML/d (2.8 mgd), with a peak capacity of 61 ML/d (16 mgd). The facility receives flow from the service area in and surrounding the city of Greencastle and discharges treated effluent to Big Walnut Creek. The facility was upgraded in 2002 to increase reliability during dry weather and also to accommodate higher peak wet weather flows. Increases in wet weather flow capacity were intended to reduce overflows in the upstream collection system. An overview schematic of the water resource recovery facility is shown in Figure 13.7.

During dry weather conditions, the influent is screened using fine screens and degritted through stacked tray grit removal systems. Screened and degritted influent is fed to the vertical loop reactor (VLR) Tank No. 1. Typically, the three tanks are run in series and are run at low dissolved oxygen levels (i.e., aerated anoxic conditions) to optimize energy and sludge settleability. Mixed liquor is settled in secondary clarifiers before UV disinfection and effluent aeration through both cascade and fine bubble diffusers, and then discharged to Walnut Creek. Under typical conditions, the mixed liquor concentration in the VLR tanks is between 4000 and 5000 mg/L. Two clarifiers are typically in operation.

Operation of the facility during wet weather includes the use of step feed/contact stabilization to store biomass in VLR Tanks No. 1 and 2. As flows to the facility increase, the screened and degritted flow is first sent to

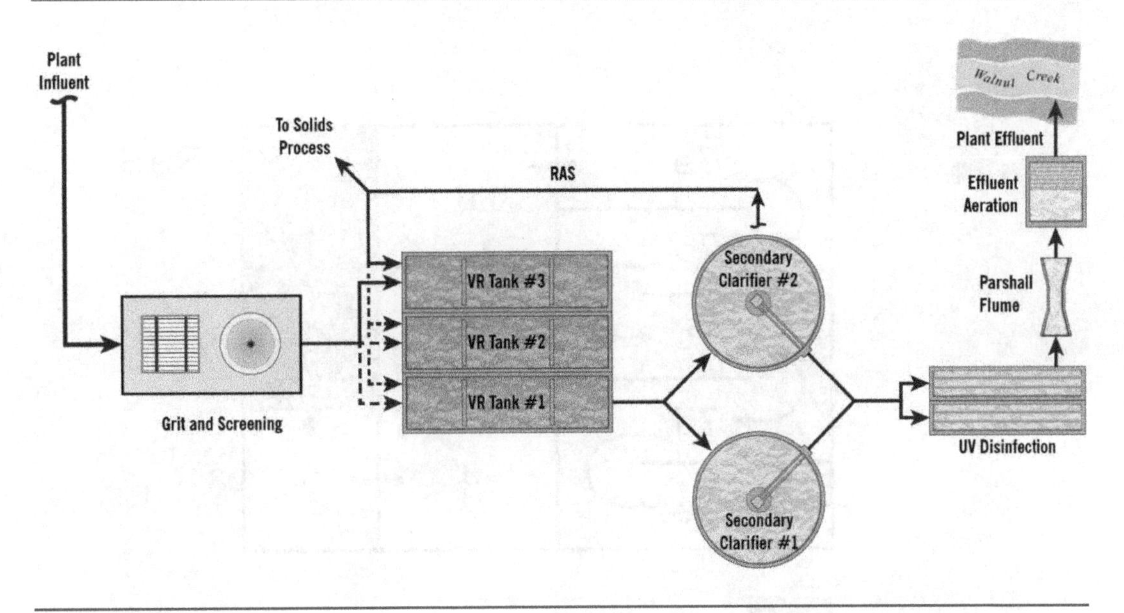

FIGURE 13.7 Overview schematic of the Greencastle, Indiana, Wastewater Treatment Plant.

VLR Tank No. 2 and solids are allowed to accumulate under mixed and aerated conditions (surface aerator) in VLR Tank No. 1. If the flows remain high for an extended period (duration depends on the peak flows and loadings during the event), influent feed can be switched to VLR Tank No. 3 and biomass is allowed to accumulate in VLR Tanks 1 and 2. The facility staff can also waste more solids from the clarifier to available solids storage tanks to remove biomass from the aeration basins during extended peak flow conditions (Gellner et al., 2012).

Changes in influent feed can also be used as a wet weather operation strategy in multistage biological nutrient removal (BNR) systems. Influent is often fed downstream of upstream nutrient removal zones and RAS is returned and accumulated in these upfront zones. An example of this strategy at the Durham Advanced Wastewater Treatment Facility (Clean Water Services, Tigard, Oregon) is shown in Figure 13.8 (Johnson et al., 2007).

3.3 Return Activated Sludge Storage

3.3.1 Return Activated Sludge Storage within Existing Reactors

Using the step feed/contact stabilization process configuration, biomass is concentrated in the upstream portions of the biological reactor during wet weather. When RAS storage is used, biomass is removed from the main process tanks during wet weather. This strategy can achieve the same reduction in mass loading to the clarifiers as with step feed/contact stabilization.

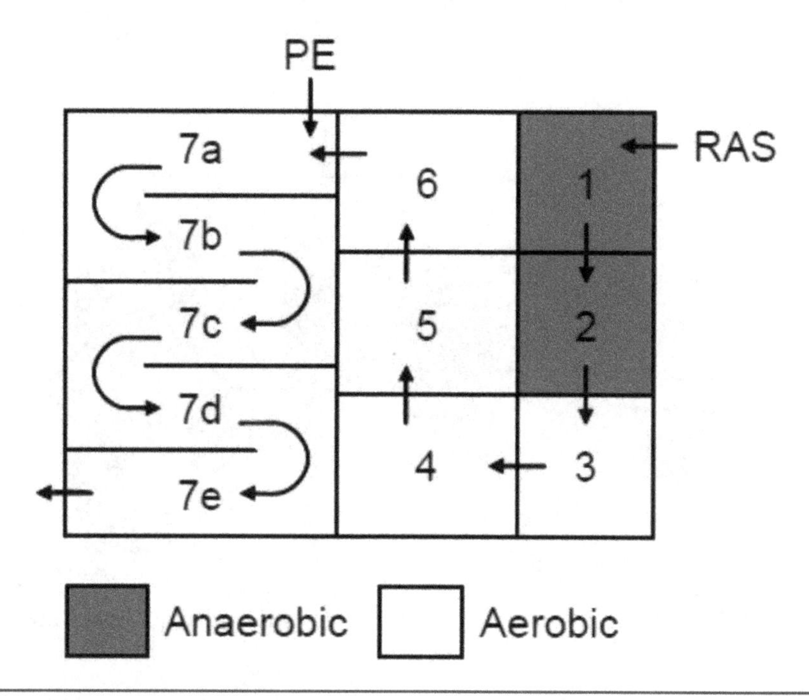

FIGURE 13.8 Example of wet weather step feed at a BNR facility. Note: Primary effluent (PE) is typically fed to Cells 1 or 3 during lower flow conditions (Johnson et al., 2007).

Return activated sludge storage refers to operation in which discrete biological treatment trains are taken offline and wet weather flow is sent to either a fewer number of biological reactors than used during dry weather or to an offline bioreactor. The RAS continues with equal splits of biomass to both online and "offline" reactors. A portion of the biomass is stored in the offline tanks during the storm event. After the storm event, influent wastewater and RAS feed return to typical operation. A schematic of this strategy is shown in Figure 13.9.

The Sycamore Creek Wastewater Treatment Plant in Cincinnati, Ohio, owned and operated by the Metropolitan Sewer District of Greater Cincinnati (MSDGC), uses RAS storage within existing reactors during wet weather to optimize facility performance. A schematic view of the facility is shown in Figure 13.10.

During typical dry weather flows, screened and degritted flow is pumped to the anaerobic reactors. Effluent from the anoxic zones is returned through internal recycle to the head of the anaerobic reactors at a rate of approximately 30.3 ML/d (8 mgd) (with all aeration tanks in service). Effluent from the anaerobic reactors is split to the four aeration tanks. The first two zones of each aeration tank are operated in anoxic mode. Flow is then discharged to the aerobic zones. A total of six different aeration zones are

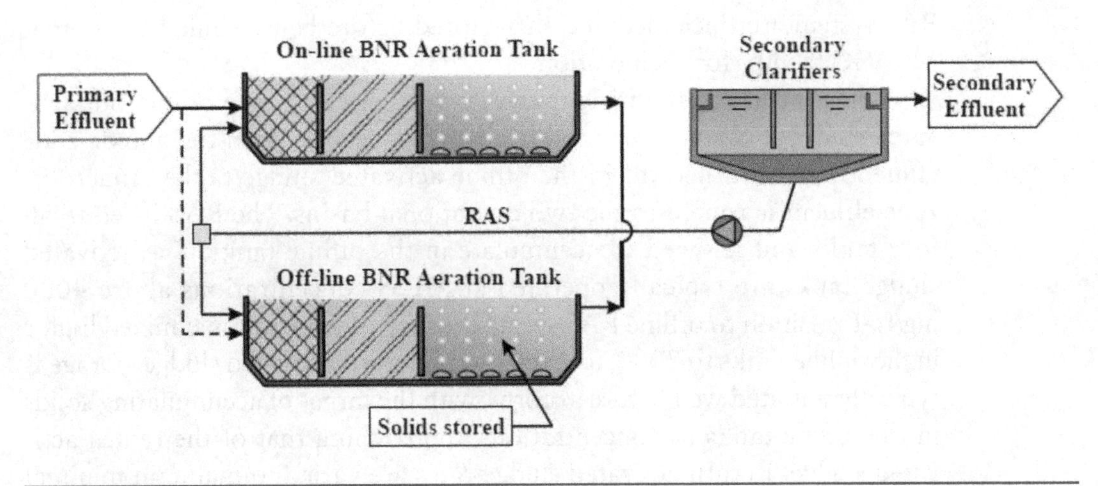

FIGURE 13.9 Schematic of an offline bioreactor used for RAS storage during wet weather (Pitt et al., 2007).

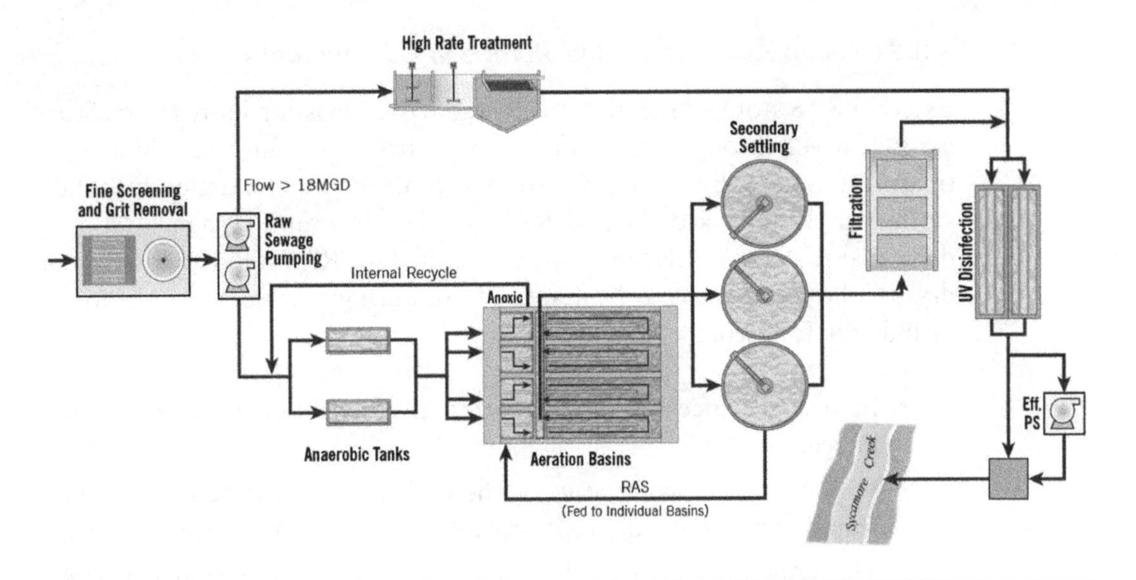

FIGURE 13.10 Simplified liquid schematic of the Sycamore Creek Wastewater Treatment Plant (Gellner et al., 2012).

included in each tank. Air is delivered to the basins through fine bubble diffused aeration.

Typically, the facility operates two final clarifiers. Individual RAS pumping for each clarifier is used to return biomass to the head of the anoxic zones at a rate of approximately 40 to 50% of influent. Secondary effluent is discharged to effluent disc filters for further polishing. The disc filters are backwashed automatically based on filter headloss. Wasted solids from the

RAS system are thickened and then stored before being hauled to another MSDGC facility for incineration.

When facility staff anticipate a wet weather event, two of the four activated sludge tanks are taken offline. Internal recycle from the anoxic zone effluent is also turned off in the offline activated sludge tanks. Anaerobic zone effluent is routed to the two operational basins. The RAS is fed to all four tanks and allowed to accumulate in the offline tanks. The activated sludge tanks are typically operated at MLSS concentrations above 4000 mg/L. Transition to offline RAS storage reduces the operational mixed liquor in the online tanks to 2000 to 2500 mg/L. Return activated sludge storage is typically initiated well before a storm, with the target of accumulating solids in the offline tanks to concentrations approaching that of the return activated sludge. Return activated sludge rates are varied to maintain minimal sludge blankets in the clarifiers. Chemically enhanced high-rate treatment is used as a supplementary parallel treatment system for influent flows that exceed 68.1 ML/d (18 mgd).

3.3.2 *Return Activated Sludge Storage in External Tanks*

External RAS storage is similar to storage within existing bioreactors. Storage is achieved through use of tanks separate from the mainstream biological treatment system. External RAS storage could be in an existing biosolids storage tank or in a dedicated RAS equalization tank. When there is a dedicated RAS tank, this tank can also be used for RAS stabilization during dry weather operations. Advantages of use during dry weather operations include the following:

- Increased process solids retention time without a corresponding increase in MLSS,
- Volume advantage relative to the mainstream bioreactor—biomass stored in the RAS equalization tank will be at RAS concentrations,
- The RAS tank can be used for sidestream treatment of sludge processing returns, and
- Improved biomass settling properties.

3.4 Changes in Biological Nutrient Removal Operating Conditions

An example of step feed/contact stabilization at a facility practicing enhanced biological phosphorus removal was presented in a previous section. There are several variations of step feed/contact stabilization that can be considered for facilities that have stringent nutrient permit levels.

Where clarifiers have adequate capacity to accommodate increases in flow and solids loading, additional strategies for nutrient removal focus on maximizing detention time for nitrogen removal and preserving anaerobic conditions for phosphorus removal. Strategies include the following:

- Partial bypass of influent around the anaerobic zone while continuing to return RAS to the anaerobic zone—this strategy minimizes the effect on detention time in the anaerobic zone and continues to maintain anaerobic/aerobic cycling to maintain conditions required for PAO growth and maintenance. When more influent is bypassed around the anaerobic zone, influent carbon is delivered directly to the downstream anoxic and aerobic zones. One downside of this approach is that phosphorus release may be slowed or impaired resulting from lack of carbon in the upfront anaerobic zones. However, phosphorus release typically returns after feed is returned to the anaerobic zones; and
- Decrease in internal recycles used for nitrogen removal—during dry weather, internal recycles high in nitrate are operated at high rates (i.e., 150 to 300%) relative to incoming wastewater flows. During wet weather events, the internal recycle rates can be reduced to increase effective detention time in the anoxic portions of the aeration basins. Where anoxic zones have been retrofitted as swing zones with diffusers to operate aerobically under wet weather conditions, some of the downstream portions of the primary anoxic zone can be operated with air to enhance nitrification. In this strategy, the level of nitrogen removal achieved in the system may be lower, resulting from lowered internal recycle and less detention time in the aerobic zones of the aeration tank.

In facilities in which secondary clarifiers have inadequate capacity for high solids loadings during peak wet weather periods, the strategies that can be used to achieve optimum system performance are more similar to step feed/contact stabilization. One difference is that a small stream of influent wastewater is sometimes maintained to the anaerobic zone to maintain anaerobic conditions and to drive the cyclic phosphorus release and uptake cycle.

3.5 Use of Settling Aids

Effective settling in the secondary clarifiers is critical to preventing biomass washout and preserving effluent quality during wet weather events. Depending on the biological system and its operation, settling can sometimes become challenging during dry weather conditions. The causes and options for addressing settling issues have been well documented elsewhere.

If poor settling conditions are present in the mainstream treatment process when a wet weather event occurs, the control of the biological process is more difficult. Sludge blankets increase in height, RAS concentrations decrease, and active management of solids between clarifiers and the biological system becomes more difficult. In facilities where poor settling has occurred or where clarifier loadings are anticipated to be beyond acceptable levels, it is possible to add coagulants or flocculants upstream of the secondary clarifiers to improve settling properties. Use of chemicals is not intended to be continuous, rather, to be applied only during times of poor settling and/or high wet weather flows. Examples of the most commonly used coagulants are ferric chloride, alum, sodium aluminate, and polyaluminum chloride. In facilities where chemical precipitation is used for phosphorus, the precipitation of phosphates often improves settling. Polymer is the most common method of aiding settling on an as-needed basis. Jar testing should be conducted to evaluate the potential benefit of chemical addition for improvement of activated sludge settling.

3.6 Mainstream Ballast

Bio-Mag (Siemens) uses oxidized iron particles before secondary clarification or added to the mixed liquor to improve the settling properties of the activated sludge. The specific gravity of magnetite is 5.2, which, when added to the mixed liquor, significantly increases the density of the MLSS floc. As a result of this increase, mixed liquor concentrations in the aeration tanks can be significantly increased without adverse effects on secondary clarifier performance. Likewise, hydraulic and solids loading rates to the secondary clarifiers can be significantly increased beyond conventional levels. Allowable loading rates to secondary clarifiers are 4 times the generally accepted practice (U.S. EPA, 2013).

Magnetite is separated from waste activated sludge through a shearing process, followed by accumulation of magnetite on a magnetic drum screen system. Depending on the design criteria used for average daily flows and loads, a treatment facility using Bio-Mag can accommodate significant increases in wet weather flow. Step feed is also possible if desired or needed to reduce clarifier loadings. Testing should be performed to verify target secondary clarifier loading levels during peak wet weather loading conditions.

3.7 Aeration Tank Settling

Aeration tank settling is a term used for turning off a portion of the air supply system in an activated sludge tank to settle solids before the secondary clarifiers and reduce solids loading. Although this practice has limited application, it can be used as a "last resort" operational practice during extended or especially severe wet weather events. When diffusers are turned off, mixed liquor solids settle in the aeration tanks and are retained until the diffusers are turned back on. Typically, aeration tank settling should be used for short periods. When extended periods of use are required, some "bumping" of process air should be used to limit mixed liquor buildup in the system (WEF et al., 2010).

3.8 Parallel High-Rate Biological Treatment

Parallel high-rate biological treatment refers to the use of a separate treatment train for flows exceeding the mainstream biological system capacity at a WRRF. When flows exceed the capacity of the mainstream biological system, a parallel biological treatment system is brought online to accommodate excess flow. A high-rate biological contact reactor upstream of a ballasted flocculation settling system can be used for high-rate biological treatment in parallel with the main treatment trains. BioActiflo (Veolia) capitalizes on the compact ballasted settling system used for high-rate physical–chemical treatment. Biological treatment is achieved through a short period of contact, followed by settling of biological solids through the ballasted flocculation process. A schematic of this system is shown in Figure 13.11.

The BioActiflo system has been piloted in several locations, including a recent demonstration pilot in Akron, Ohio. The first full-scale installation of the technology was commissioned in 2013 at the Wilson Creek

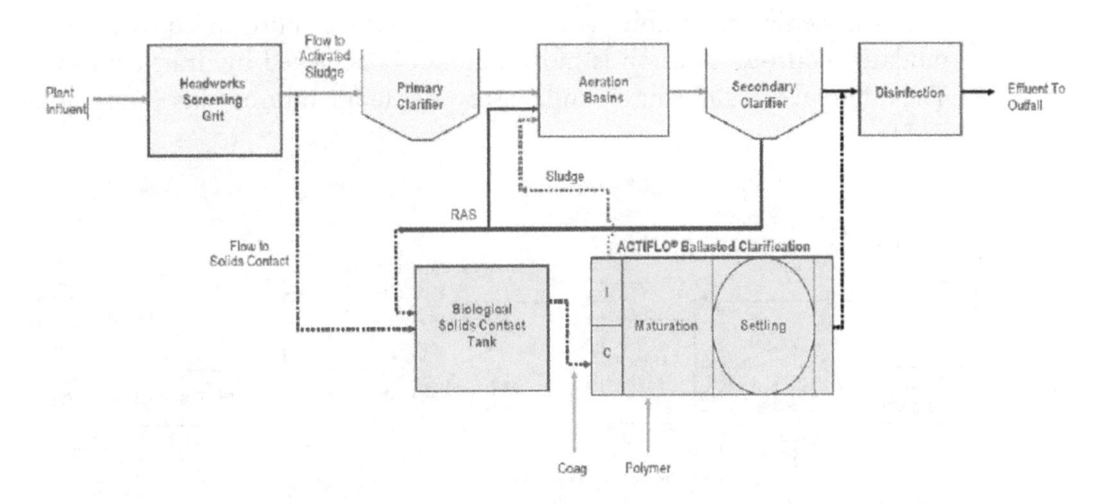

FIGURE 13.11 Schematic of a typical configuration of BioActiflo™ for wet weather treatment (U.S. EPA, 2012).

Regional Wastewater Treatment Plant. In this application, the high-rate physical/chemical process is used during dry weather for tertiary polishing of phosphorus to levels less than 0.1 mg/L. During wet weather events, a portion of the biological system is used for a short detention time biological contact process, followed by settling in the high-rate process (Sandino et al., 2011). A schematic of the treatment process used at the Wilson Creek facility is shown in Figure 13.12.

3.9 Fixed Film Processes

Fixed film processes incorporate the use of a biofilm system growing on a fixed or floating media for biological treatment. Fixed film technologies commonly used by utilities for biological treatment include the following:

- Trickling filters,
- Integrated fixed film activated sludge (IFAS),
- Moving bed biofilm reactor (MBBR), and
- Biologically active filter (BAF).

3.9.1 Trickling Filters

Tickling filters have been used for decades for cost-effective removal of carbonaceous BOD. Some trickling filter systems are also used to nitrify. Trickling filters use an even distribution of influent wastewater across the top of a media, preferably with a large amount of pore space to accommodate aeration. Aeration is most commonly achieved by draft aeration. The difference in ambient air and wastewater temperatures causes a natural

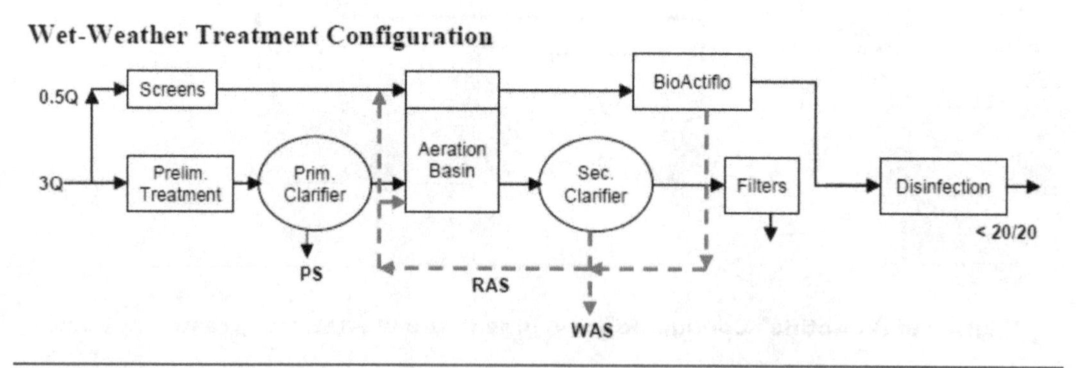

FIGURE 13.12 Wet weather flow schematic for BioActiflo™ at the Wilson Creek facility (Sandino et al., 2011).

draft, resulting from temperature-induced air currents. In the summer, the direction of draft is up and in the winter, it is down.

Trickling filters are operated with a relatively high recycle, typically 2 to 3 times the incoming wastewater, to control wetting rates on the media and stable operation of the attached biomass (WEF et al., 2010). Where trickling filters are used as the main biological treatment for a facility, the following strategies are viable alternatives:

- Decrease recirculation rates so that total influent flow to trickling filters remains the same. Using this strategy, the recirculation rates from the recirculation pump station would be reduced as more flows come into the facility and even turned off if influent flows increase beyond the typical recirculation rates. This strategy would be expected in facilities where trickling filters are used as the main biological removal system; and
- Diversion of flow to other process units—when flows increase to rates greater than base flow and recirculation or where the facility uses trickling filters for roughing, it may be necessary to divert some flows around the trickling filters. If biological treatment is necessary, there are two options: a parallel treatment biological system, such as BioActiflo, or, if solids contact is used downstream, flow could be diverted to this reactor for high-rate biological treatment before secondary clarification.

3.9.2 Integrated Fixed Film Activated Sludge

Integrated fixed film activated sludge uses a combination of suspended growth and fixed biofilm growth in the same reactor. This system provides greater capacity in a similar footprint, through use of both fixed and suspended-phase biomass. Fixed biomass is retained on media, either floating or fixed. For floating media, the media is held within the tank through the use of submerged screens. For fixed media, the biomass is held on media supported on a fixed frame (WEF et al., 2010).

When flows increase in IFAS systems, the key consideration is the hydraulic capacity of the aeration tanks and the secondary clarifiers. For floating media systems, the designer should also verify that increased forward velocity in the aeration basins does not adversely affect media distribution and that it does not exceed allowable headloss through the submerged media retention screens. Because fixed media systems have stationary media, distribution of media is not a concern. If secondary clarifiers have been designed for higher hydraulic loading rates, it may be possible to accommodate the entire peak flow through the IFAS reactors. Similar to conventional systems, step feed/contact stabilization could also be used for solids loading reduction to the secondary clarifiers.

3.9.3 Moving Bed Biofilm Reactor

The moving bed biofilm reactor includes biomass retained on media for biological treatment. The key difference between MBBR and IFAS systems is that MBBRs do not have a return line from the subsequent solids/liquid separation stage, resulting in low levels of suspended growth population providing treatment. Suspended biomass is made up of sloughed/eroded biomass from the media. Settled solids are captured downstream of the MBBR with a clarifier or other high-rate alternative technology. As with IFAS systems, the media is retained in the tank through submerged screens (floating media) or by a support frame (fixed media) (WEF et al., 2010).

Similar to IFAS systems, the most important factors in effectively treating wet weather flows are the hydraulic capacity of the aeration tank, the approach velocity to the media retention screens (if applicable), and the capacity of the clarifier/solids separation technology. If the solids separation system has the ability to accommodate the additional flows, then continuing to feed at the head of the aeration tanks is permissible. If the separation system/technology does not have sufficient capacity for the wet weather flows, then a parallel treatment train is likely the best alternative for the process. Removal efficiencies of an IFAS or MBBR system during wet weather will be less than during dry weather because of reduced detention times in the biological system.

3.9.4 Biological Active Filter

Biological active filters combine filtration with an attached biofilm removal of wastewater constituents. Biological active filters use an engineered media for filtration and biofilm growth and retention. Biological active filters can be used as the main biological process for BOD and/or nitrification or as a tertiary process for nitrogen removal. If BAFs are used for denitrification, the BAFs are used as part of the tertiary treatment processes. The main advantage of the process is its small footprint, especially when used for the main biological treatment step in the facility. Because the unit process also includes filtration, downstream clarifiers are not required.

Biological active filter capacity is based on the filtration rate (flow per unit of area) and constituent loading rate. In facilities where a BAF is the main treatment source, BAFs should be sized for expected peak capacities. If expected peak flows are higher than those capacities, the alternatives for increasing wet weather capacity include construction of additional process units or a parallel high-rate treatment system.

Where BAFs are used for tertiary nitrogen removal in denitrification filters, high flowrates can be accommodated by increasing secondary effluent to the capacity of the BAF unit process and then diverting the remaining portions of the flow around the BAF system to the treatment facility effluent. Although lower removal efficiency of nitrogen can be expected under this scenario, this reduction can be coordinated with facility effluent permit limits, depending on the level to which nitrogen must be removed.

3.10 Membrane Processes

Membrane processes incorporate the use of membranes for separation of solids from the biological reactor effluent in lieu of secondary clarifiers. In membrane systems, hollow fiber, or plate-style membranes, are placed in a dedicated tank and suction or gravity is used to achieve membrane flux and production of treated effluent. The advantage of membrane systems is the high effluent quality resulting from small pore size and increased MLSS concentration in the upstream biological tank. In many instances, concentrations up to 8000 to 10 000 mg/L can be maintained in the biological system (WEF et al., 2010).

One limitation of membrane systems is the need to operate below critical flux rates for effluent production. The flowrate for membrane systems is proportional to the area of membranes supplied. Therefore, the capital cost for a membrane system to treat all flows is typically restrictive for situations in which peak flowrates are high. Typical membrane treatment systems have a practical peaking factor of 2 times the average influent flow. Upstream flow equalization is often included in the design for higher influent flows. In locations where higher secondary treatment capacity is required, alternatives for providing a parallel high-rate biological system should be evaluated. For this application, a portion of the MLSS from the membrane bioreactor could be diverted to a high-rate contact basin, followed by conventional solids separation.

4.0 DESIGN CONSIDERATIONS

The design of treatment systems for wet weather handling takes advantage of key process concepts that have been documented elsewhere in the design of WRRFs (WEF et al., 2010). Management of wet weather events requires close coordination of operation before and during wet weather events. This section will provide simple examples of considerations to be included in the comprehensive planning and design of strategies to accommodate wet weather. Careful consideration of design flexibility, operations, and system capacities should be used for each specific system. Each is unique and should be evaluated carefully.

4.1 Influent Characterization

4.1.1 Flow and Loads

Before final design, available historical data on influent wastewater properties and changes associated with wet weather events should be assessed. The expected variations in peak flow and load should be quantified for different frequency storm events. The most important criteria for wet weather characterization include the following:

- Magnitude and duration of the first flush—the headworks and associated preliminary treatment and primary treatment system must be adequately designed for the potential for first flush. Increases in TSS, inert solids, particulate COD, and soluble COD can also occur. The first flush magnitude and duration depends on storm intensity, collection system characteristics, and time since the last rain event. Biological treatment systems should be designed to accommodate potential short-term increases in TSS and BOD loading from first flush events; and
- Increases in loadings resulting from a wet weather event—often, after passage of the first flush, loadings are more representative of maximum month or maximum week conditions. Therefore, in the absence of additional data, wet weather loadings that exist for extended periods could be made by assuming maximum flows and maximum month or peak week loadings. The designer should use caution in these assumptions and make sure that site-specific conditions are considered.

4.1.2 Level of Service/Duration

Where possible, the level of service should be defined as part of the design criteria. Typically, the wastewater flows and resulting hydrograph at the WRRF are tied to the level of service defined for the collection system. For separate systems, the level of service is typically defined by the return frequency storm used for design. For combined systems, the level of service is defined by target numbers of overflows within a collection system or at the WRRF. Where possible, design storms and associated hydrographs for the desired level of service should be established.

4.2 Reduced Performance Resulting from Decreased Detention Time

During design of a wet weather management system, careful consideration should be given to reduced removal efficiencies through treatment units. Reduced detention times and lower amounts of biomass in the system (resulting from operation in step feed/contact stabilization or RAS storage, for instance) will result in lower removal efficiencies for most wastewater constituents. Careful consideration should be given to the effect of wet weather events on mass-based discharge permit limits at a treatment facility. Reduced performance during the wet weather event will often necessitate operation at enhanced levels during the remaining permit period.

4.3 Contact/Bioflocculation Time

When step feed/contact stabilization is used as a wet weather strategy, the influent wastewater is fed downstream from conventional entry points along the aeration tank. Feed at downstream points allows for accumulation of solids in the upstream zones of the bioreactor and reduced solids loading to the secondary clarifiers.

Where feasible, it is recommend that provisions for multiple step feed points be incorporated. Depending on the configuration of the aeration basins, two to three points are recommended. The cost of multiple feed points should be weighed against the benefit of flexibility in operations. Having multiple feed points allows for adjustment in strategy, particularly with short duration storms that would not require extended periods of modified wet weather operation. For those storms, feed at an intermediate point further upstream would preserve process integrity. For longer-duration storms, use of the downstream feed point may be necessary, particularly if the zones upstream of the first feed point contain biomass concentrations approaching the RAS concentration.

Minimum retention times are required for bioflocculation for effective wastewater solids capture and rapid COD uptake. There is no readily accepted design guidance for minimum effective flocculation time. Sandino et al. (2011) provided a minimum of 15 minutes of biological contact time for the BioActiflo system at the Wilson Creek facility. Jimenez et al. (2005) found that removal of particulate constituents and colloidal material in high-rate biological contact systems varied with MLSS and detention time. Esping et al. (2012) used bench-scale testing to quantify levels of removal under different conditions and to determine optimum sizing. Designers should consider field sampling and jar testing analysis to confirm that the sizing of the contact zone is sufficient for bioflocculation and that BOD and TSS removal are sufficient for effluent discharge permit requirements. A similar approach was used by Esping et al. (2012) in the evaluation of wet weather strategies at the South Shore Water Reclamation Facility in Milwaukee, Wisconsin.

4.4 Aeration

Adequate air must be supplied for biological treatment processes and to maintain sludge settleability. There is not a unique requirement for aeration during wet weather operations. Designers should provide sufficient air to maintain minimum operating dissolved oxygen in aerobic zones and in areas of solids reaeration (i.e., upstream portion of the aeration tank during step feed/contact stabilization). The designer should coordinate process air needs in these areas with typical operation and make provisions for additional headers and diffusers in areas where step feed/changes in feed pattern will create additional oxygen demand during wet weather operations. For RAS equalization/storage, sufficient air must be supplied to maintain aerobic conditions in those tanks to maintain biological activity and viability when dry weather conditions return.

4.5 Clarifiers

Design practices and conventions for secondary clarifiers have been well documented elsewhere (WEF, 2005; WEF et al., 2010). This section will focus on consideration of circular secondary clarifier features, which aid in effective management of wet weather flows.

4.5.1 Existing Capacity

Secondary clarifiers are most often designed to criteria similar to 10 States Standards (GLUMRB, 2004). Strategies for wet weather typically involve consideration of additional clarifiers to provide the hydraulic and solids settling capacity or by considering higher loading rates than typical criteria for significant increases in facility flow. Evaluation of capacity is influenced by the level of risk a utility is willing to accept for operation relatively infrequently in a given year.

The ability to model the secondary clarifiers by computational fluid dynamics (CFD), allows the designer to evaluate capacity on a dynamic basis and to weigh the risk factors associated with wet weather flow conditions. Analysis by CFD also allows consideration of the ability of the facility infrastructure to accommodate relatively short-duration high-flow peaks vs long sustained storm events. Information from CFD can also be used to optimize the physical layout/configuration of the weirs, solids density baffle, and flocculating center well.

When clarifier capacity requires that a clarifier be operated at loading levels higher than the 10 States Standards, it is strongly recommended that field stress testing and CFD analysis be considered to help identify the risk of failure and the potential boundaries for wet weather operation. The higher the loading rates, the more crucial the appropriate application of a CFD model to the evaluation of the risk. It is also recommended that the designer consider additional features for the clarifiers if they are not equipped with the appropriate components.

4.5.2 Sidewater Depth

Clarifiers with sidewall depths greater than 3.7 m (12 ft) are recommended where new facility infrastructure is required for wet weather operations. Higher sidewall heights (i.e., up to 5.2 to 5.5 m [17 to 18 ft]) have been used in multiple utilities across the country for enhanced performance during wet weather. Higher sidewater depths provide additional solids storage during wet weather operation and the ability to deal with changes in solids loadings during these transitions. The designer should evaluate the potential benefit of this additional storage against the cost of deeper construction.

4.5.3 Energy Dissipating Inlets

Energy dissipating inlets (EDIs) on circular clarifiers dissipate the energy of the incoming mixed liquor and encourage flocculation of mixed liquor solids. Where modifications are being made to existing clarifiers or new clarifiers are required, the incorporation of EDIs is strongly encouraged to optimize wet weather performance.

4.5.4 Optimized Diameter and Depth of Center Well

The dimensions of the flocculating center well can have important implications in performance during wet weather. Typical design criteria for flocculating center wells include 20 minutes of detention time and/or 30 to 35% of clarifier diameter. The depth of the center well skirt is also important. If it is too deep, erosion of settled sludge blankets can occur. If it is too shallow, currents from the EDI can go below the skirt and affect settling in the quiescent clarification zone. In addition, if the depth of the well skirt is too deep, currents can erode settled sludge blankets (WEF, 2005). Computational fluid dynamics modeling can help to evaluate optimum sizing of the center well through simulation of a variety of loading characteristics, from dry weather to wet weather. The dimensions yielding the best estimated effluent suspended solids are typically implemented, unless cost is prohibitive.

4.5.5 Return Activated Sludge Pumping Capacity

Conventional activated sludge systems are typically designed for RAS rates of 20 to 100% of annual average design flow (WEF, 2005). Higher RAS rates may be necessary for wet weather management, particularly when step feed/contact stabilization or RAS storage is used. Higher return flowrates increase the rate of solids removal from the clarifier during high solids loading conditions when transitions are being made to operation in a wet weather mode. During these periods, increases in sludge blanket depth will occur when solids loading rates exceed the solids mass returned in the RAS system. Dynamic modeling is a useful tool in the evaluation of required solids return rates during wet weather.

The type of solids removal and its relation to pumping is also an important consideration. Quick withdrawal of solids is sometimes necessary for managing sludge blankets and solids inventory. Therefore, systems with direct piped suction discharge to the pumping system are preferred for wet weather over sludge handling systems that involve wet wells or telescoping valves. Systems in which individual clarifiers can be isolated to individual pumps with discrete flow meters on each line are ideal for solids management.

4.5.6 Influence of Solids Settleability

The ability to effectively settle solids is essential to stable and effective biological treatment. When a facility experiences poor settleability of solids, typically quantified by high sludge volume index (SVI), removal of solids in the clarifiers becomes less effective. In addition, process stability is also affected because of dilute sludge blankets. The solids separation issues associated with poor solids settleability can be exacerbated in wet weather. The effective clarifier capacity is diminished when solids have higher SVIs and clarifier performance is further limited by higher hydraulic loading and solids loading rates that often occur during wet weather. When considering wet weather improvements, the designer should also consider historical sludge settleability. Where SVIs are chronically high or where the facility experiences a wide range of SVIs, anaerobic/anoxic selectors should be considered as part of the improvements. Selectors will help to ensure effective settling during dry weather and, therefore, help control poor settling effects during wet weather. Where selectors are not practical, provisions for addition of chemical settling aids during wet weather events should be included.

4.6 Solids Inventory Management

4.6.1 Introduction

Facility operations staff must be able to effectively monitor and control sludge blankets during transition into, during, and after a wet weather event. Where step feed/contact stabilization or RAS storage are used, solids are moved from clarifiers to a biological system to adjust clarifier solids loadings and to maintain biomass within the system. Attention must be given in design to the transitions to and from wet weather operation. During these times, there will be transients in solids loading that must be effectively managed to prevent solids washout from the system.

A simplified example of step feed/contact stabilization is used in this section to illustrate the dynamic behavior of biological systems during wet weather. Simple mass balance calculations are used to approximate solids location and loadings at different times during dry/wet weather operation.

4.6.2 Step Feed/Contact Stabilization Example

In this example, the WRRF experiences an annual average flowrate of 3.8 ML/d (1 mgd) and operates a conventional four-pass plug flow aeration basin system with circular clarifiers. Total volume of the aeration basin system is 3.8 ML (1 mil. gal) and the total surface area of the clarifiers is 232 m^2 (2500 sq ft) (Miklos, 2013). A schematic of the system operated during wet weather is shown in Figure 13.13.

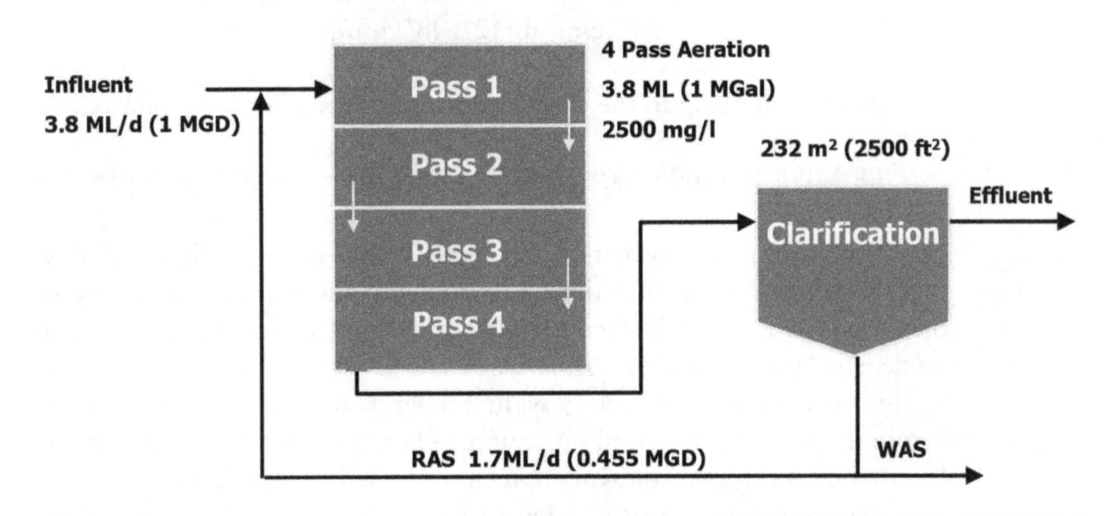

FIGURE 13.13 Four-pass activated sludge system during dry weather.

Additional assumptions for this example include the following:

- Dry weather MLSS = 2500 mg/L,
- Effective settling with SVI = 120 mL/g, and
- RAS concentration = 8000 mg/L.

Return activated sludge rates required to keep solids blankets at consistent levels can be calculated by simplified mixing formulas, assuming no wasting and negligible solids in the effluent, using the following equation:

$$\text{RAS flow for equilibrium} = (\text{Influent } Q \text{ (ML/d)} \times \text{MLSS} \times \text{(mg/L)} / (\text{RAS} \times \text{(mg/L)} - \text{MLSS} \times \text{(mg/L)}) \qquad (13.1)$$

The calculated RAS rate for equilibrium conditions is

$$\text{RAS flow for equilibrium} = (3.8 \text{ ML/d} \times 2500 \text{ mg/L}) / (8000 \text{ mg/L} - 2500 \text{ mg/L}) = 1.7 \text{ ML/d } (0.455 \text{ mgd})$$

The operating solids loading to the secondary clarifiers is as follows:

$$2500 \text{ mg/L} \times (3.8 \text{ ML/d} + 1.7 \text{ ML/d}) / 24 \text{ h} = 573 \text{ kg/h } (1264 \text{ lb/h})$$

Solids loading on an area basis to the clarifier is

$$2500 \text{ mg/L} \times (3.8 \text{ ML/d} + 1.7 \text{ ML/d}) / 232 \text{ m}^2 = 59 \text{ kg/m}^2\cdot\text{d } (12.1 \text{ lb/sq ft/d})$$

Hydraulic loading to the clarifier during dry weather is as follows:

$$3.8 \text{ ML/d} \times 1000 / 232 \text{ m}^2/24 \text{ h} = 0.68 \text{ m}^3/\text{m}^2\cdot\text{h } (400 \text{ gpd/sq ft})$$

During wet weather events, the example WRRF must handle a peak flow of 11.4 ML/d (3 mgd) or 3 times the annual average flow. A schematic of the flow scheme when the facility switches to step feed/contact stabilization mode is shown in Figure 13.14.

In this example, the facility waits for flows to increase to 11.4 ML/d before moving into step feed operation. When flows reach 11.4 ML/d, the facility opens the gate, allowing influent to be fed to Pass 3, and closes the gate that allows influent to be fed to Pass 1. The solids loading initially increases because the incoming flow is conveying undiluted MLSS to the

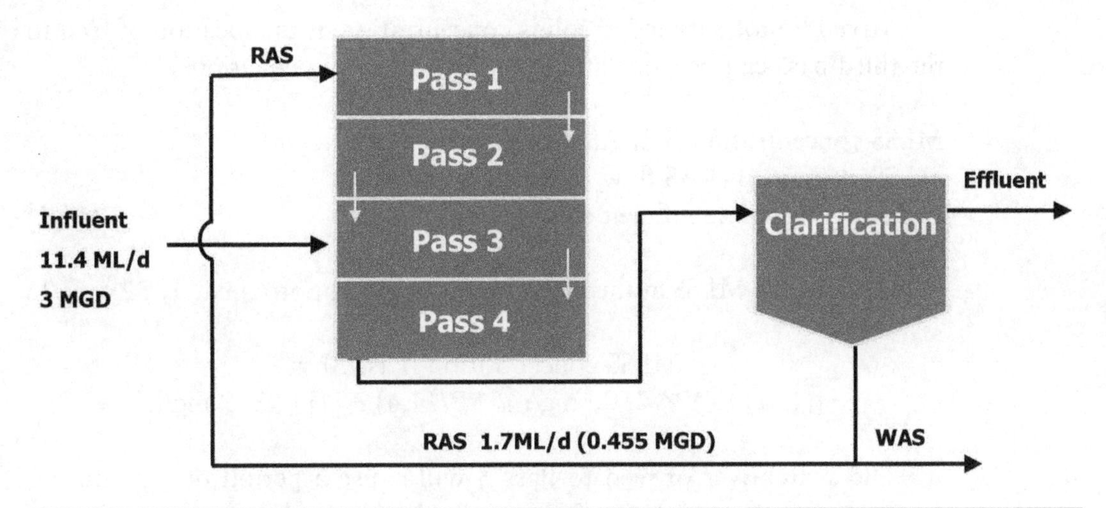

FIGURE 13.14 Four-pass activated sludge system operated in step feed/contact stabilization (initiation of step feed).

clarifiers. At the point where flows have increased to 11.4 ML/d, solids loading to the clarifiers increases by approximately 3 times to

$$\text{Clarifier solids loading} = 2500 \text{ mg/L} \times (11.4 \text{ ML/d} + 1.7 \text{ ML/d})/24 \text{ hours} = 1364 \text{ kg/h } (3000 \text{ lb/h})$$

$$\text{Clarifier solids loading} = 2500 \text{ mg/L} \times (11.4 \text{ ML/d} + 1.7 \text{ ML/d})/232 \text{ m}^2 = 141 \text{ kg/m}^2\cdot\text{d } (29 \text{ lb/sq ft/d})$$

Physically, solids are being sent to the clarifiers faster than they are being withdrawn. The withdrawal rate, using the original concentration of the RAS, is 573 kg/h (1265 lb/hr). For a short period during initiation of wet weather, more solids will be sent to the clarifiers than are withdrawn through the RAS unless return rates are increased. Sludge blankets in the clarifiers would begin to rise until the solids in the RAS equal the solids loading rate to the clarifiers or until the RAS rate is increased. The concentration in the RAS would need to increase to approximately 19 200 mg/L to equal the solids loading rate of approximately 1364 kg/h (so RAS = 1.7 ML/d at 19 200 mg/L). Alternately, RAS rates would need to be increased to 4.1 ML/d at 8000 mg/L to prevent increases in sludge blankets. If there is sufficient sidewater depth in the clarifier to accommodate the increased sludge blanket depths for short periods, no adjustments of the RAS would be needed.

Mixed liquor suspended solids concentration at the location of feed to the third pass can be calculated using the following equation:

MLSS concentration = [RAS flow (fraction of influent) × MLSS × (mg/L)]/(RAS flow (fraction of influent) + 1.0), assuming negligible influent solids (13.2)

Initially, the MLSS at the point of feed will be approximately 329 mg/L.

$$\text{MLSS concentration (Pass 3)} = [(1.7/11.4) \times 2500\ \text{mg/L})/(1.7/11.4) + 1] = 329\ \text{mg/L}$$

The switchover of feed to Pass 3 will cause a period of time during which the clarifier solids loading actually drops to values corresponding to this lower MLSS of 329 mg/L. When this plug of flow reaches the end of the activated sludge basin, solids loading to the clarifiers would be reduced to approximately 180 kg/h (396 lb/hr) or 18.6 kg/m^2·d (3.8 lb/sq ft/d):

$$\text{Clarifier solids loading} = 329\ \text{mg/L} \times (11.4\ \text{ML/d} + 1.7\ \text{ML/d})/24\ \text{hours} = 180\ \text{kg/h (396 lb/hr)}$$

$$\text{Clarifier solids loading} = 329\ \text{mg/L} \times (11.4\ \text{ML/d} + 1.7\ \text{ML/d})/232\ \text{m}^2 = 18.6\ \text{kg/m}^2\cdot\text{d (3.8 lb/sq ft/d)}$$

Sludge blanket levels will begin to drop or return rates must be decreased to maintain a sludge blanket level in the clarifiers (see Figure 13.15).

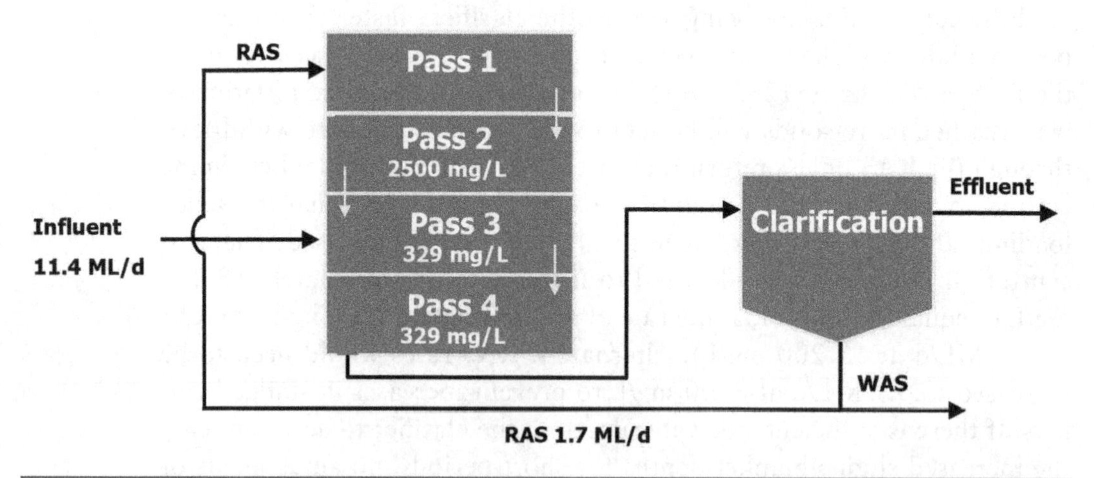

FIGURE 13.15 Four-pass activated sludge system operated in step feed/contact stabilization (time = 4 hours).

If wet weather flows continue for an extended period, the biomass concentration in the first two passes will continue to approach RAS concentrations. Note that depending on the level of solids in the secondary clarifiers before the storm event, RAS concentrations may change over time and RAS rates may need to be adjusted to manage sludge blanket levels. Operations staff must monitor solids inventory carefully throughout the activated sludge system to maintain effective treatment.

4.6.3 Transition Considerations

The example presented in the previous section is a simplified example of changes in solids location and distribution that occur during changes in operation for wet weather through the biological system. The transition into and out of wet weather operation in the biological system is extremely important. Designers should evaluate potential strategies and operation during the design process.

When wet weather events can be anticipated, it is most effective to begin transition to wet weather operations before the wet weather event. In the example presented, the increase in solids loading that resulted from the change of feed to Pass 3 could have been minimized if the transition in feed location had occurred at lower flows. Operations staff at the Sycamore Wastewater Treatment Plant typically begin transition to wet weather operation a full 12 hours before the storm event (Gellner et al., 2012).

Designers should also anticipate the need to carefully transition back to dry weather operation after the storm event. This transition can take significant periods of time depending on the mode of wet weather operation used. Sandino et al. (2011) used modeling to evaluate proposed transitions between dry weather and wet weather and found that transition back to typical dry weather operations may require 7 to 10 days because of increased solids inventory generated during wet weather.

4.7 Modeling for Design of Wet Weather Systems

Use of WRRF process models is increasingly common practice for designers. When possible, it is recommended that a properly calibrated process model be used to evaluate and refine design criteria and process unit sizing. The process model developed should be calibrated using available data from the facility and supplemental data gathered specifically for model development. Details on acceptable levels of model calibration and recommended calibration practices are included elsewhere (International Water Association, 2013).

Process models allow for efficient consideration of multiple scenarios and various dynamic simulations. When wet weather improvements are being considered, process models are useful to evaluate the following:

- Process performance during storm events—where information on influent hydrographs is available or where these can be simulated using a collection system model, use of dynamic simulations can help the designer evaluate performance of the proposed wet weather treatment under dynamic conditions. In many cases, the highest peak flows occur over short periods (i.e., 4 to 12 hours). In these situations, process units may accommodate loading rates much higher than are typical with acceptable effluent quality;
- Effluent permit compliance—the mass of constituents discharged during wet weather events is a significant concern when WRRFs have strict effluent requirements based on mass for a stipulated permit compliance period. Predicted performance during wet weather can be used to determine the feasibility of meeting permit requirements after a wet weather event for various scenarios; and
- Modifications to operations—the transition into and out of wet weather operations can create peak loading situations, as was described in the simplified example. Modeling can help identify those periods of high transient loading and can also help to evaluate modifications to operations to mitigate the process effects of these loadings and associated process performance. Transitions back to dry weather operation can also be evaluated and optimized.

5.0 REFERENCES

Esping, D.; Krill, B.; Parker, D.; Jimenez, J.; Fitzpatrick, J.; Yang, F.; Bate, T. (2012) Comparison of Three Wet Weather Flow Treatment Alternatives to Increase Plant Capacity. *Proceedings of the 85th Annual Water Environment Federation Technical Exhibition and Conference* [CD-ROM]; New Orleans, Louisiana, Sept 29–Oct 3; Water Environment Federation: Alexandria, Virginia.

Fortin, J.; Pitt, P.; O'Connor, P.; Giardina, F.; Husband, J.; Koch, C. (2007) Newtown Creek Demonstration Project Wet Weather Operating Strategies. *Proceedings of the 80th Annual Water Environment Federation Technical Exhibition and Conference* [CD-ROM]; San Diego, California, Oct 13–17; Water Environment Federation: Alexandria, Virginia.

Gellner, W.; Miklos, D.; Tabor, C.; Sandino, J.; Fitzpatrick, J.; Bradley, P.; Ott, D.; Jenkins, A. (2012) What Happens When Everyone "Flushes" at the

Same Time—Case Studies on Management of Wet Weather Flow Events. *Proceedings of the 85th Annual Water Environment Federation Technical Exhibition and Conference* [CD-ROM]; New Orleans, Louisiana, Sept 29–Oct 3; Water Environment Federation: Alexandria, Virginia.

Great Lakes Upper Mississippi River Board of State and Provincial Public Health and Environmental Managers (2004) *Recommended Standards for Wastewater Facilities,* 2004 ed.; Health Research, Inc.: Albany, New York.

International Water Association (2013) *Guidelines for Using Activated Sludge Models;* Scientific and Technical Report No. 22; IWA Publishing: London, U.K.

Jimenez, J.; Parker, D.; Bratby, J.; Schuler, P.; Campanella, K.; Freedman, S. (2005) In the Absence of the Blending Policy: A Novel High Rate Biological Treatment Process. *Proceedings of the 78th Annual Water Environment Federation Technical Exhibition and Conference* [CD-ROM]; Washington, D.C., Oct 29–Nov 2; Water Environment Federation: Alexandria, Virginia.

Johnson, B.; Mengelkoch, M.; Baur, R. (2007) Full Scale High Rate Wet Weather Biological Contact Performance. *Proceedings of the 80th Annual Water Environment Federation Technical Exhibition and Conference* [CD-ROM]; San Diego, California, Oct 13–17; Water Environment Federation: Alexandria, Virginia.

Katehis, D.; Sandino, J; Daigger, G. (2011) Maximizing Wet Weather Treatment Capacity of Nutrient Removal Facilities. *Proceedings of the 84th Annual Water Environment Federation Technical Exhibition and Conference* [CD-ROM]; Los Angeles, California, Oct 15–19; Water Environment Federation: Alexandria, Virginia.

Miklos, D. (2013) *Wet Weather Operation. Optimizing Your Secondary Treatment Process;* Presentation Given to Central Ohio Utilities, Columbus, Ohio.

Pitt, P.; van Niekerk, A.; Garrett, J.; Hildebrand, L.; Bailey, W. (2007) Management of Wet Weather Storm Flow Conditions at BNR/ENR Treatment Plants. *Proceedings of the IWA/WEF Nutrient Removal Specialty Conference;* Water Environment Federation: Alexandria, Virginia.

Sandino, J.; Covington, J.; Kraemer, J.; Boe, R.; McKnight, D.; Cole, B. (2011) First of Its Kind Full-Scale Implementation of a Biological and Chemically Enhanced High-Rate Clarification Solution for the Treatment of Wet Weather Flows at a Municipal Wastewater Treatment Facility. *Proceedings of the 84th Annual Water Environment Federation Technical Exhibition and Conference* [CD-ROM]; Los Angeles, California, Oct 15–19; Water Environment Federation: Alexandria, Virginia.

U.S. Environmental Protection Agency (2013) *Emerging Technologies for Wastewater Treatment and In-Plant Wet Weather Management*; EPA-

832/R/12011; U.S. Environmental Protection Agency, Office of Wastewater Management: Washington, D.C.

Wang, L.; Shammas, N.; Hung,Y., Eds. (2009) *Advanced Biological Treatment Processes;* Humana Press: New York.

Water Environment Federation (2005) *Clarifier Design,* 2nd ed.; WEF Manual of Practice No. FD-8; Water Environment Federation: Alexandria, Virginia.

Water Environment Federation; American Society of Civil Engineers; Environmental & Water Resources Institute (2010) *Design of Municipal Wastewater Treatment Plants,* 5th ed.; WEF Manual of Practice No. 8; ASCE Manuals and Reports on Engineering Practice No. 76; Water Environment Federation: Alexandria, Virginia.

14

High-Rate Treatment

James D. Fitzpatrick, P.E.; Matt Crow, P.E.; Thomas A. Lyon, P.E.; and C. Robert O'Bryan, Jr., P.E.

1.0 INTRODUCTION

In many instances, a significant portion of the wet weather flows to a water resource recovery facility (WRRF)—particularly those with high peak flowrates and low pollutant concentrations—can be most effectively and efficiently treated by physical and chemical means. Physical and chemical clarification processes have been used to improve water quality since the dawn of civilization (Baker and Taras, 1981). Clarification processes (i.e., sedimentation, flotation, and filtration) are widely used in WRRFs and, since about the 1970s, many technologies relying on these mechanisms have been adapted and optimized for the treatment of wet weather flows.

The treatment profession has typically used total suspended solids (TSS) removal efficiency along with hydraulic loading rate to describe the performance of clarification technologies. This publication uses the term, *high-rate treatment (HRT),* to mean technologies that provide TSS removal equivalent to at least primary clarification, but at significantly higher hydraulic loading rates than typically practiced for conventional dry weather applications. As discussed further in Chapter 12, the influent fraction of nonsettleable TSS (TSS_{non}) limits the effluent quality capable by gravimetric separation alone. Thus, the effluent quality from HRT alternatives such as retention treatment basins, vortex separators, and lamella settlers (without chemical enhancements) would typically be expected to be similar to that from conventional primary clarifiers and are discussed further in Chapter 12.

Enhanced high-rate treatment (EHRT) is used to further distinguish those technologies that can operate at the high hydraulic loading rates of HRT, but provide significantly higher TSS removal. These performance classifications are illustrated in Figure 14.1, showing typical ranges for HRT and EHRT technologies in comparison to the ranges typically observed for conventional and chemically enhanced primary clarifiers. The comparatively higher TSS removals attained by EHRT technologies are typically ascribed to mechanisms that either alter the influent particle settling characteristics (i.e., coagulation, flocculation, and floc ballasting) or physically remove a larger portion of TSS_{non} (i.e., filtration). The remainder of this chapter focuses on these EHRT technologies, many of which are also discussed in guidance and technology transfer documents by the U.S. Environmental Protection Agency (U.S. EPA) (2013) and the Water Environment Federation (2013).

2.0 PROCESSES AND TECHNOLOGIES

Since the mid-1990s, significant advances have been made in clarification technologies tailored for treating wet weather flows. The Water Environment

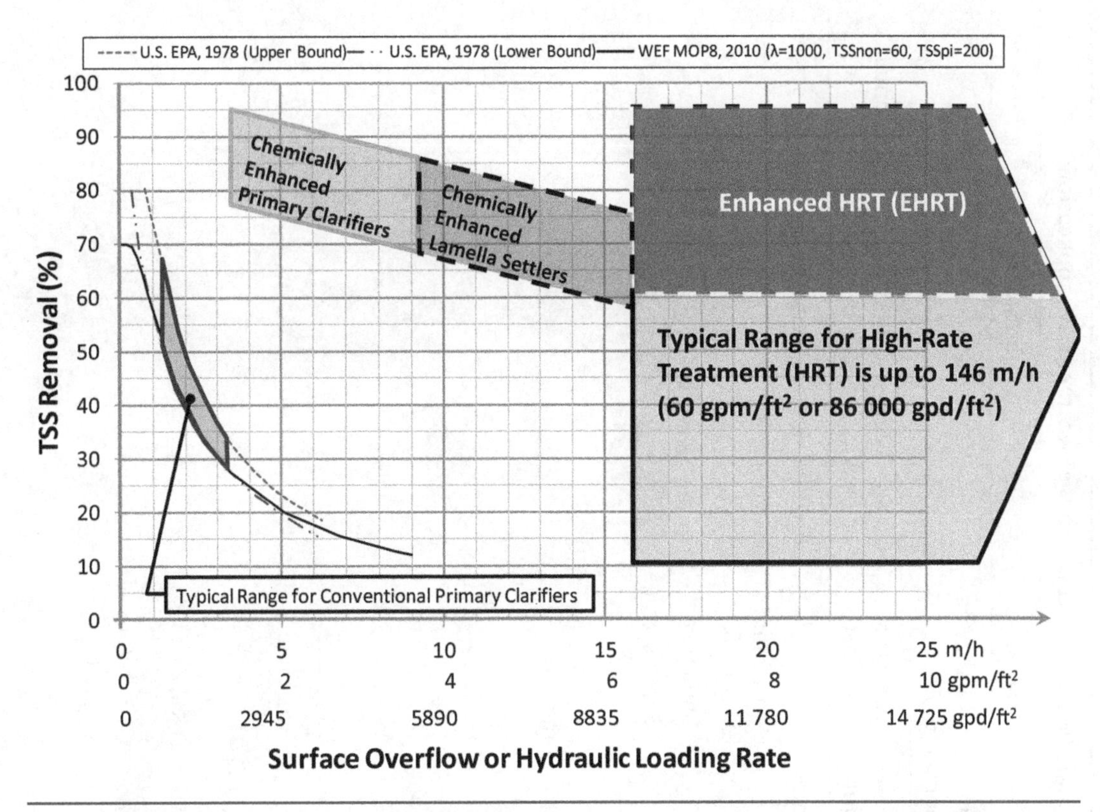

FIGURE 14.1 Typical performance ranges of advanced clarification processes compared to conventional gravity separation (i.e., primary clarification).

Research Foundation (WERF) (2002) provides good documentation of many of these developments, particularly for those technologies that rely on chemical enhancements combined with lamella settlers. More recently, significant advances have also been made in filtration technologies. These advances have been mainly focused on increasing the hydraulic efficiency of these process units while also improving effluent quality to enhance the reliability of effluent disinfection processes (see Chapter 15). The following subsections provide narrative descriptions of these technologies, and Section 3.0 and Table 14.1 provide further details and design criteria.

2.1 Chemically Enhanced Sedimentation Alternatives

Chemically enhanced sedimentation (CES) alternatives build upon the basic coagulation and flocculation mechanisms described in Chapter 12 for chemically enhanced primary treatment (CEPT). The terms, *CEPT* and *CES*, are sometimes used interchangeably in the profession; however, for this specialized publication, *CEPT* will refer specifically to the use of conventional primary clarifier basins that are also used for primary treatment during dry

TABLE 14.1 Typical ranges of design criteria for selected physical and chemical processes in wet weather flow applications.

Process unit and design criteria	Chemically enhanced sedimentation			High-rate filtration		
	Coagulation and flocculation	Solids contact and ballasted flocculation				
		Sludge recirculation	External ballast	Compressible media	Cloth media	Floating media
Preliminary treatment requirements[a]	Coarse screen	12 to 20 mm screen	Microsand, 6 to 10 mm, 2D screen; Magnetite, 12-mm screen	Downflow, coarse screen; Upflow, 12-mm screen	Coarse screen	Coarse screen
Coagulant addition[b]						
Chemical	Ferric chloride, ferric sulfate, aluminum sulfate, polyaluminum chloride			[c]	[c]	[d]
Dose, mg/L as Fe^{3+} or Al^{3+}	5 to 25	5 to 25	15 to 50			
Mixing intensity	G = 50 to 500 s^{-1}	G = 200 to 300 s^{-1}	G = 200 to 300 s^{-1}			
Mixing time, seconds	30 to 900	60 to 120	60 to 120			
Flocculant addition[b]						
Chemical	Anionic polyacrylamide (aPAM)			[c]	[c]	[d]
Dose, mg/L as active polymer	0.2 to 1.5	0.5 to 3	0.5 to 2			
Mixing intensity	G = 200 to 300 s^{-1}	G = 200 to 300 s^{-1}	G = 200 to 300 s^{-1}			

Mixing time, seconds		30 to 60	60 to 180	60 to 180			
Flocculation[b]							
Mixing intensity		G = 20 to 50 s^{-1}	G = 100 to 200 s^{-1}	G = 100 to 200 s^{-1}	[c]	[c]	[d]
Mixing time, minutes		3 to 20	3 to 5	3 to 5			
Liquid/solid separation							
Hydraulic loading rate	m/h	3.4 to 9.3	37 to 98	73 to 146	24 to 29	20 to 24	20 to 42
	gpm/sq ft	1.4 to 3.8	15 to 40	30 to 60	10 to 12	8 to 12	8 to 17
Basin sidewater depth	m	3.0 to 4.6	4.6 to 9.1	4.6 to 9.1	1.8 to 4.3	2.4 to 3.0	3.0 to 4.6
	ft	10 to 15	15 to 30	15 to 30	6 to 14	8 to 10	10 to 15
Ballast recovery and makeup							
Ballast recovery method		N/A[h]	N/A	Microsand, hydrocyclone; Magnetite, magnetic drum	N/A	N/A	N/A
Ratio to treated effluent,		0.5 to 1	0.5 to 1	4 to 6	2 to 10	2 to 10	2 to 10
Thickness, % total solids		2 to 4	2 to 4	0.1 to 0.5	0.03 to 0.1	0.03 to 0.1	0.03 to 0.1

(*continued*)

TABLE 14.1 Typical ranges of design criteria for selected physical and chemical processes in wet weather flow applications (*continued*).

Process unit and design criteria	Chemically enhanced sedimentation			High-rate filtration		
	Coagulation and flocculation	Solids contact and ballasted flocculation				
		Sludge recirculation	External ballast	Compressible media	Cloth media	Floating media
Ancillary support systems						
Coagulant storage and feed	[e]	[e]	[e]	N/A	N/A	[d,e]
Flocculant storage and feed	[e]	[e]	[e]	N/A	N/A	[d,e]
Ballast storage and feed	N/A	N/A	[e]	N/A	N/A	N/A
Sludge/backwash pumping	Required	Required	Required	Required	Required	Required
Backwash blowers	N/A	N/A	N/A	[f]	N/A	N/A
Effluent TSS, mg/L[b, g]	15 to 40	15 to 40	15 to 40	15 to 40	15 to 40	15 to 40[d]

[a]Site-specific design or operational constraints may require other alternatives to be considered. Refer to Chapter 11 for additional details.

[b]Assumes influent water chemistry, particle size distribution, and settling characteristics that have been typically observed for municipal wet weather flows. Assumes that influent contains adequate alkalinity and no substances that inhibit chemical dose responses typically observed. Site-specific tests are recommended to confirm dose response for each application.

[c]Chemical coagulation and flocculation not typically required for filtration alternatives, but may be used for removal of additional TSS, phosphorus, or certain other materials if required by site-specific water quality requirements.

[d]Additional demonstration testing recommended to verify performance reported by system manufacturers. Hattori (2008) and Hayashi (2009) reported that no coagulants or flocculants were required, but Yoon et al. (2012) reported that alum and polymer were required to achieve effluent TSS in the range shown here, but at a different site with different media.

[e]Refer to Chapter 7 and Section 3.4 of this chapter for additional details regarding chemical and ballast storage alternatives.

[f]Air scour requires 0.05 m^3/s of air per square meter of media bed surface (10 scfm/sq ft).

[g]Assumes influent contains TSS in the range of 50 to 150 mg/L.

[h]N/A = not applicable.

weather conditions, whereas *CES* will refer to applications with nonconventional basin designs used in an auxiliary fashion to intermittently increase capacity during wet weather flow events.

2.1.1 Chemically Enhanced Sedimentation with Lamella Settlers

Besides conventional basin designs, chemical coagulation and flocculation have also been applied upstream of lamella settler basins to decrease basin footprint requirements in at least one WRRF in North America (Liyanage, 2010) and are fairly standard alternatives for industrial process wastewater clarification. Crow et al. (2012) reported that pilot tests in King County, Washington, demonstrated TSS removals greater than 50% with settling tank hydraulic loading rates of up to approximately 34 m/h (20 000 gpd/sq ft or 14 gpm/sq ft).

2.1.2 Solids Contact and Ballasted Flocculation

The solids contact process builds on the basic CES coagulation and flocculation steps, but flocculated solids from the clarifier sludge blanket are recirculated to the zone, where polymer is added and floc formation begins. A generalized process flow diagram of these adaptations to the basic CES process is shown in Figure 14.2, and Chapter 12 of *Design of Municipal*

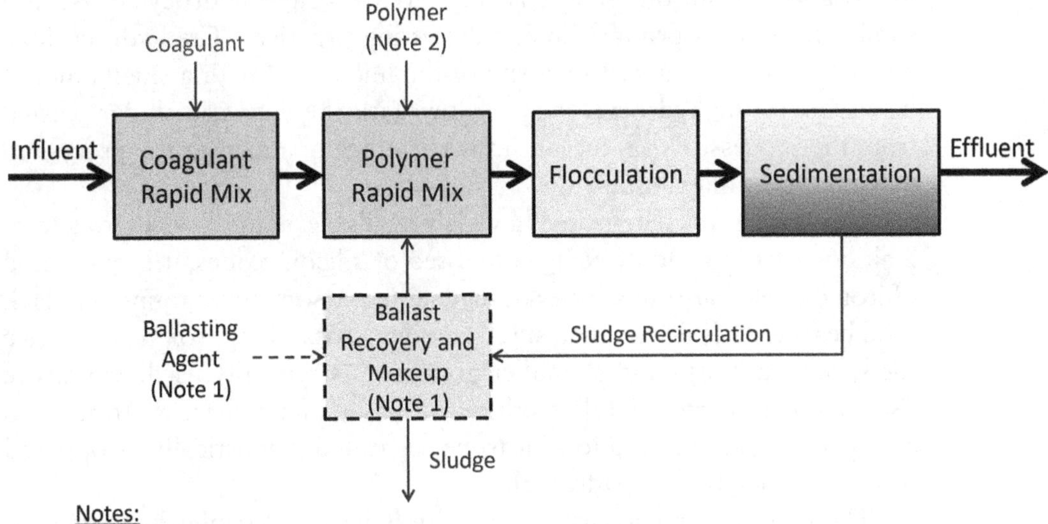

FIGURE 14.2 Process flow diagram for chemically enhanced sedimentation with solids contact and ballasted flocculation.

Wastewater Treatment Plants (WEF et al., 2010) provides further descriptions of these alternatives as well. The recirculated solids act as nucleation sites and help establish a "critical mass" of solids to achieve the "sweep flocculation" thought to maximize the removal of colloidal material. The sludge recirculation also helps increase floc density and integrity with thickened sludge from the bottom of the tank acting somewhat like a ballast material for lighter influent particles and floc. Some alternatives accomplish this additional solids contact in a "reactor" tank that is separate from the settling tank itself, while others incorporate recirculation equipment within the clarifier basin itself in a solids contact zone separated by baffles from the settling zone.

Besides sludge recirculation, some alternatives also incorporate an external ballast material as part of the flocculation process. This external ballast material helps to further increase floc density and settling velocities, thus decreasing the required settling area. Coagulation and flocculation are the primary mechanisms responsible for converting TSS_{non} to large particles that are more easily settled; therefore, these basic CES steps are the biggest factors in determining the effluent quality from ballasted flocculation CES processes. Increased floc density and the use of lamella settlers are the primary process components that decrease settling basin footprint requirements.

Small-grained silica sand (microsand) and magnetite are two external ballast alternatives most commonly used in WRRF applications. In the microsand option, the ballast is recovered through hydrocyclones, in a similar fashion as primary sludge degritting practices. The hydrocyclone underflow contains a vast majority of the microsand with a small amount being lost in the hydrocyclone overflow with the degritted sludge wasted from the process. Fresh microsand is periodically added to the process to make up for ballast losses.

Like microsand, magnetite is a widely available, naturally occurring mineral; however, instead of being composed of silicon oxides, it is composed of iron oxides. Large quantities of magnetite are sometimes found on black sand beaches. It is denser than silica-based microsand (approximately twice the specific gravity) and a smaller grain size is typically used. Magnetite itself is not a magnet, but, like other iron-based materials, it is attracted to a magnetic field, which allows it to be separated magnetically as opposed to gravimetrically (i.e., hydrocyclones).

The main differences with magnetite ballasted flocculation compared to microsand ballasted flocculation are that magnetite has the potential to produce floc with slightly higher settling velocities (i.e., slightly less settling area is required), and slightly higher ballast recovery efficiencies are feasible with magnetic separation as opposed to gravimetric separation. The magnetic ballast recovery equipment used in this alternative was adapted from separation processes used in the mining industry.

Full-scale solids contact and ballasted flocculation facilities have been operating in Europe, North America, and Asia for more than a decade. Their performance in full-scale wet weather applications has also been documented by Crow and Coxon (2011), Fitzpatrick et al. (2008), Keller et al. (2005), and many others. Many proprietary systems are now commercially available such as DensaDeg and CONTRAFAST (sludge recirculation), ACTIFLO and RapiSand (microsand ballasted flocculation), and CoMag (magnetite ballasted flocculation).

2.2 High-Rate Filtration Alternatives

Besides sedimentation, advancements have also been made in applying filtration alternatives to wet weather flows. Although conventional granular media filters have been used to treat water for centuries and are also commonly used for tertiary treatment of WRRF effluents, they are somewhat limited in their feasibility for wet weather flows because of their relatively low solids loading capacity and resulting backwash requirements. Deep-bed filters offer some advantages in that regard, but are still somewhat limited by the relatively high headloss characteristics of conventional granular media. Design and operational considerations for tertiary applications with granular media filters and some of the other alternatives discussed further herein are described in Chapter 16 of *Design of Municipal Wastewater Treatment Plants* (WEF et al., 2010). Many of the considerations described there also apply to wet weather flows, with the significant technical differences for wet weather flows being the solids loading rate and particle size distribution of the solids, which are key process criteria for sizing of filtration units.

2.2.1 Compressible Media

Compressible media filtration (CMF) has been used for more than a decade in full-scale wet weather treatment applications. In the 1990s, the technology was part of a full-scale combined sewer overflow (CSO) technology demonstration program implemented by the Columbus Water Works (Columbus, Georgia), assisted by a grant from U.S. EPA with their Office of Research and Development serving as a quality assurance reviewer and WERF serving as a peer reviewer (WERF, 2003). Oij et al. (2010) describe stream health improvements since the 2007 startup of a CMF-based stormwater treatment facility on Weracoba Creek, an urban stream in Columbus, Georgia.

Compressible media filtration uses a bed of synthetic fiber balls to capture influent suspended solids and is currently available as either the Fuzzy

Filter or the WWETCO FlexFilter. In the Fuzzy Filter, the media bed is compressed mechanically in the direction opposite of the bulk liquid flow (i.e., countercurrent) between a fixed perforated plate on the inlet side of the bed and a movable plate on the outlet side of the bed. As illustrated in Figure 14.3, the Fuzzy Filter is most commonly configured as an upflow filter with influent flowing up through the media bed, but it can also be configured in a downflow arrangement. The filter remains in filtration mode until the captured solids accumulate to the point that the media must be cleaned. At that point, a wash cycle is initiated and the movable plate of the bed is moved to allow the bed to expand. An air scouring backwash process is used to clean the media and the solids are carried away in the wash water stream. At the end of the wash cycle, the media is recompressed and the unit is returned to service after the remaining solids are flushed from the system. The Fuzzy Filter can also be configured without mechanical compression. This alternate configuration has the advantage of no moving parts, but also does not have any means to increase bed compression beyond that provided in the typical co-current direction by fluid hydraulics.

The WWETCO FlexFilter also uses synthetic fiber balls in its media bed, but has a slightly different operating cycle than the Fuzzy Filter. It uses a downflow configuration, and the influent hydrostatic head is used to compress the bed transversely to the direction of fluid flow between flexible, reinforced membranes that form a conically shaped bed profile. This

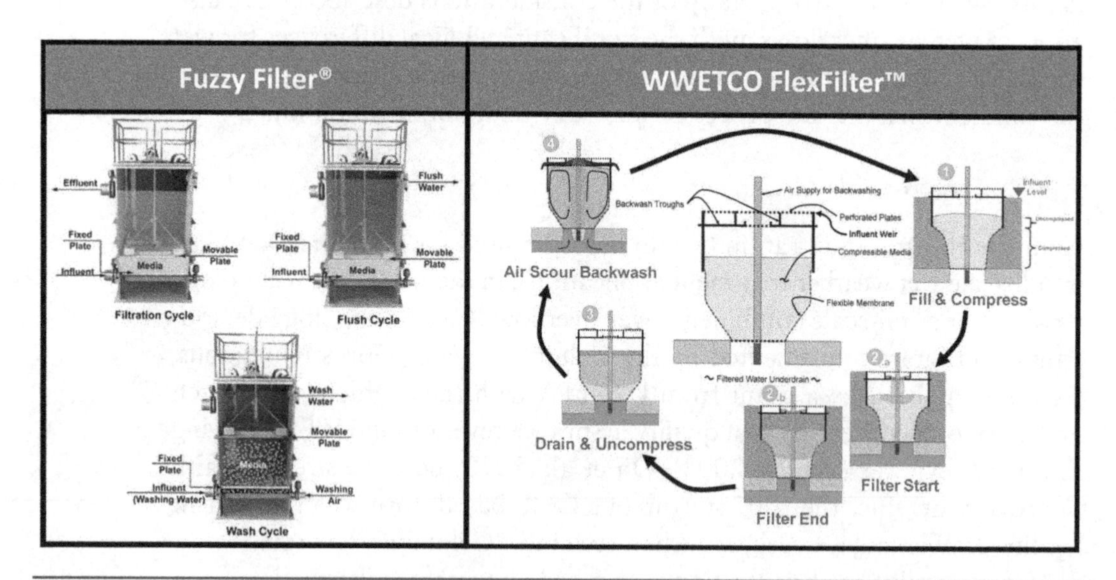

FIGURE 14.3 Two variations of compressible media filtration (courtesy of Schreiber LLC and WesTech Engineering, Inc.).

transverse media compression is in addition to the typical co-current compression provided by the bulk fluid flow. The WWETCO FlexFilter also has a slightly different air scouring arrangement. As illustrated in Figure 14.3, the lower portion of the WWETCO FlexFilter is a reinforced flexible membrane manufactured from an engineered fabric designed for the required operating head of the system. As influent fills the basin around the filter housing, the membrane flexes inward and compresses the media. Influent continues to rise until it overtops the influent weir and flows onto the media bed. As solids accumulate on and within the filter bed, the influent level over the media bed rises to the point that signals the start of a backwash cycle. The basin is then drained to release the compression and the media is backwashed. An air scour pipe along the centerline of the filter housing cleans the media, and the solids are lifted by the air and carried away in a backwash trough.

For continuous-flow applications, multiple CMF cells must be provided so that individual cells can be cleaned using a backwash system while the remaining cells continue operating in the filtration mode. The CMF process is typically operated as a constant-rate process (i.e., a flowrate relatively constant with increasing headloss across the filter bed until a backwash cycle is initiated). The duration of the filtration cycle is a function of the solids loading rate, and the beginning of the wash cycle is typically automated either through the use of timers or by pressure or level instrumentation monitoring the headloss across the filter bed.

2.2.2 Cloth Media

Cloth media filters are becoming increasingly common in WRRF tertiary filtration applications and offer many of the same advantages as compressible media. Cloth media filters are currently being used in conjunction with sedimentation to treat peak wet weather flows at some WRRFs in the United States and stormwater runoff in Switzerland (Baumann, 2013). Different cloth media are typically classified as being of either woven construction (i.e., relatively thin flat sheets like a microscreen) or fiber pile construction (i.e., a comparatively thicker profile similar to pile carpeting). The media is most commonly mounted on disc-shaped elements, which are configured in either an "inside-to-outside" or "outside-to-inside" flow arrangement, depending on a number of design considerations. At least one manufacturer (Aqua-Aerobic Systems) offers other geometries (drum and diamond laterals) that should also be considered, particularly for retrofits of existing drum and traveling bridge systems. Refer to Chapter 16 of *Design of Municipal Wastewater Treatment Plants* (WEF et al., 2010) for more details about cloth media technology, equipment alternatives, and diagrams of different cloth media filters.

Proper design of the cloth media is especially critical to managing the relatively higher solids loading encountered in wet weather applications compared to tertiary filtration or polishing applications. As with any filtration technology, backwash frequency is an important design consideration that is discussed further in Section 3.3.

2.2.3 Upflow Floating Media

In the early 2000s, Japan's Ministry of Land, Infrastructure, Transport and Tourism (MLIT) launched a comprehensive CSO control campaign that included research and development of a wide range of wet weather flow treatment technologies under their program called "Sewage Project, Integrated and Revolutionary Technology for the 21st Century" or "SPIRIT 21" (Horie et al., 2011). Hattori (2008) and Hayashi (2009) described a high-rate upflow filter technology that was certified by MLIT under the SPIRIT 21 program and features a floating media with a unique pinwheel shape. This technology has been retrofitted to conventional primary sedimentation tanks and used for treatment of both dry weather and wet weather flows. Typical design and performance criteria shown in Table 14.1 are based on literature from the manufacturer (Metawater). The first installation was completed in 2007 and, as of 2012, there were 25 installations of this technology with design capacities ranging up to 360 ML/d (95 mgd). A 507-ML/d (134-mgd) facility is scheduled for startup in 2014 (Kanaya, 2013). Another manufacturer, BKT United, has developed another high-rate filtration technology that uses floating spheres constructed of expanded polypropylene (instead of the pinwheel-shaped media unique to the Metawater filter) and has other somewhat different design details. Yoon et al. (2012) described demonstration-scale pilot testing of this filter (called "BBF-F") that was adapted from BKT's biofilter technology, which was first installed in full scale in 2006. The filtration-only function for CSO was verified by the Korean Ministry of Environment in 2013 under third-party testing carried out by the Korea Environmental Industry & Technology Institute. As a result of that pilot testing, a New Environmental Technology Verification Program was approved based on more than 12 months of demonstration testing at 3.8-ML/d (1-mgd) scale (Korean EPA, 2013). Full-scale installations of the BKT filtration technology are currently being constructed at the 696-ML/d (184-mgd) Seonam and 496-ML/d (131-mgd) Junrang Wastewater Treatment Plant in Seoul, South Korea. These large-scale facilities are expected to be online in 2014 (Min, 2013).

Diagrams of the Metawater high-rate filtration system and the BKT BBF-F filter are illustrated in Figure 14.4. The concept of floating media filtration has also been used in other water treatment applications besides wet weather flows. For example, buoyant polyethylene media is used in the

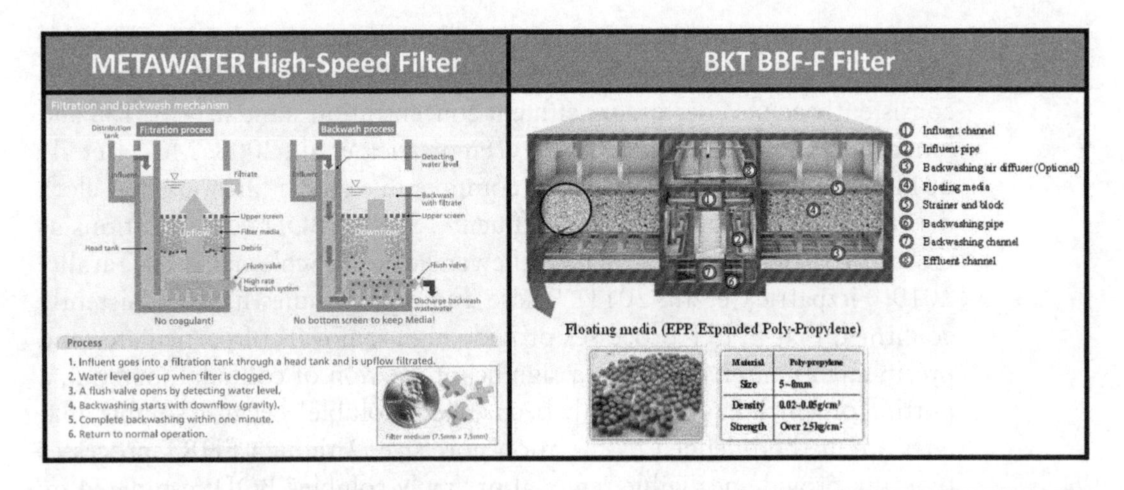

FIGURE 14.4 Two variations of upflow floating media filters (courtesy of Metawater USA, Inc., and BKT United).

adsorption clarifier for drinking water pretreatment, and a floating bed of polystyrene beads is used in the BIOSTYR biologically active filter. However, outside of the Metawater and BKT works cited previously, there have been limited installations of these technologies applied specifically to wet weather flows. As with any emerging technology, additional performance testing and demonstration studies by objective third-party researchers are recommended to help further the development of these filtration alternatives.

3.0 DESIGN CONSIDERATIONS

There are a number of factors and considerations that need to be taken into account when designing physical and chemical processes for wet weather flow treatment facilities. Typical ranges for process design criteria are summarized in Table 14.1 and discussed further in the following subsections. The focus of this publication is on wet weather facilities that are co-located at an existing WRRF. Stand-alone satellite facilities typically have other factors (remote locations, unstaffed, intermittently operated, etc.) that may require somewhat different design considerations.

3.1 Effluent Quality

As discussed in Section 1.0, EHRT processes are able to achieve lower effluent TSS concentrations than conventional primary clarification. Besides destabilizing colloids (i.e., coagulation), the metal salt coagulants used for CES also precipitate orthophosphates, resulting in significant phosphorus removal. Co-precipitation of certain metals also occurs with CES. Full-scale studies have repeatedly demonstrated the ability of parallel CES trains to consistently achieve secondary effluent concentration standards for TSS and biochemical oxygen demand (BOD) (Fitzpatrick et al., 2008; Melcer et al., 2011). Side-by-side pilot studies of compressible media filtration and cloth disc filters demonstrated similar effluent TSS and BOD concentrations as CES (Fitzpatrick, Gilpin, Kadava, Kliewer, Pekarek, Schlaman, and Tarallo, 2010; Fitzpatrick et al., 2011). These and other studies have consistently confirmed that EHRT processes provide high removals of particulates and precipitated matter, including a significant fraction of colloidal material, a portion of which has historically been called "soluble" because soluble analytes are method-defined by filtration pore size. Although EHRT processes typically provide negligible removal of "truly soluble" BOD (estimated by flocculation and filtration methods), secondary treatment standards are based on total BOD, not soluble BOD.

Some ballasted flocculation facilities have observed effluent foaming that has been attributed to surfactant pass-through; therefore, designs should minimize air entrainment in the effluent or consider defoaming agents or other measures to mitigate the potential formation of effluent foam (Fitzpatrick et al., 2005).

3.2 Process Sizing Criteria

The optimum chemicals, dosages, and mixing energies are somewhat specific for each CES application and site-specific jar testing is typically recommended as part of the design process to help identify key water chemistry constraints and process design criteria. Refer to Chapter 12 for additional design considerations for coagulation and flocculation facilities. Care should be taken in the design of polymer mixing facilities in CES alternatives to avoid entraining air into the floc matrix, which may hinder floc settling. Variable-speed mechanical flocculators often provide the most flexibility to optimize the flocculation energy over the wide range of flowrates. Lamella settlers and floc ballasting significantly decrease settling tank footprint requirements, but it is important to understand that the overall wet weather facility design must consider the footprint requirements for all required process units, not just the liquids/solids separation unit. For example, the coagulation and flocculation tanks ahead of the settling tank in a ballasted flocculation system add approximately another 170% to the system footprint; therefore, a system designed with a settling tank hydraulic loading rate of 147 m/h (60 gpm/sq ft) will have an overall system hydraulic loading rate of approximately 54 m/h (22 gpm/sq ft).

During planning studies, the footprints of upstream preliminary treatment units, downstream disinfection units, influent/effluent flow control structures (flow splitters, pumping stations, etc.), and ancillary systems such as chemical storage and feed, sludge/backwash pumping, and backwash blowers must be considered in the overall facility design as well as process unit layouts, egress, and the additional space needed for maintenance access. These planning footprints should be developed based on constructed installations of similar capacity or empirical relationships between capacity and footprint (see Chapter 5 for further considerations about facility layout).

3.3 Sludge/Backwash Management

Sludge or backwash pumps are typically used to convey the solids captured from CES and filtration alternatives to the existing WRRF for further processing. The resulting solids and hydraulic loading rates to the existing facilities should be evaluated as part of the overall facility design. For thin sludge and backwash streams (< 2% total solids), it is common practice to route these streams to the existing headworks or primary influent for co-settling and thickening with primary solids; however, separate thickening facilities can also be considered. For CES with solids contact, it is common practice to construct the sedimentation tank with additional sidewater depth to increase sludge compaction and thickening, allowing the sludge stream to be sent to residuals processing systems of the WRRF (see Chapter 16).

For high-rate filtration alternatives, proper design of the filter media and cleaning systems play a significant role in backwash performance and are critical for managing the relatively high solids loading encountered in this application compared to tertiary filtration. Media selection should be coordinated with the filter equipment manufacturer and site-specific pilot testing is typically recommended. Some filter technologies, such as compressible media filtration, use an air-scoured backwash. Blower capacity that may already be available at the existing WRRF should be considered in the design of blower facilities for such high-rate filters.

3.4 Storage and Feed of Chemicals and Ballasting Agents

Chemically enhanced sedimentation alternatives obviously require storage and feed facilities for the coagulant and flocculant used in the process. Refer to Chapter 7 for further discussion of alternatives commonly used for these ancillary systems. The wide range of influent flowrates typically encountered during wet weather events results in chemical feed turndown ratios that are somewhat higher than typically encountered in other WRRF applications.

The easier startup and shutdown of an emulsion-based polymer makeup system may outweigh the typical advantages that a powder-based system provides in more continuous operations. The CES process consumes alkalinity; therefore, sodium hydroxide or alternative coagulants may be required when treating low alkalinity waters or when the CES effluent will be discharged to downstream processes that have alkalinity constraints. In some instances, a dual coagulant process has been used with sodium aluminate or high-basicity grades of polyaluminum chloride, compensating for the alkalinity consumption from the primary coagulant.

Ballasted flocculation alternatives also require storage and feed facilities for the ballasting agents. Microsand and magnetite are stored and fed as dry powders. They are commonly handled in 20- to 40-kg (50- to 80-lb) bags and manually fed, as needed, to maintain the ballast concentration required by the process. Larger bulk bags are commonly used for initial system charging, and a bulk bag feed system should be considered for large-capacity systems.

One of the potential advantages of high-rate filtration alternatives is that chemicals are typically not required for wet weather flow applications. This potential is somewhat site-specific, depending on the particular influent characteristics (most notably, particle size distribution), the effective particle size cutoff of the filtration technology being considered, and site-specific effluent quality requirements. Therefore, site-specific studies and testing are typically recommended.

3.5 Freeze Protection

The intermittent nature of wet weather flows can make freeze protection a significant facility design issue in some climates. Some EHRT facilities circulate potable water or treated effluent through their basins at a relatively low flowrate to mitigate against freezing, algae growth, and septicity between events and reduce the ramp-up time during event startup. Others completely drain tanks and piping between events. Freeze protection is also critical for many coagulants and flocculants. The design of chemical storage tanks, piping, and equipment should consider heat tracing and insulation or locations in climate-controlled rooms.

3.6 Pretreatment and Posttreatment Requirements

Besides the core process unit, most physical and chemical treatment facilities have some sort of upstream preliminary treatment to remove relatively large debris or readily settleable solids (also known as *grit*) that could cause downstream equipment failures or maintenance problems. Effluent disinfection may also be needed depending on the bacteria standards of the receiving stream.

3.6.1 Preliminary Treatment

Refer to Chapter 11 for descriptions of screening and grit removal alternatives. For some HRT technologies, a screen opening of 25 to 80 mm (1 to 3 in.) may be adequate; however, the typical trend is to provide screens in the range of 6 to 25 mm (0.25 to 1.0 in.). Microsand ballasted flocculation typically requires two-dimensional screening down to 6 to 10 mm (0.025 to 0.0375 in.), depending on the hydrocyclone apex size. As with screenings, grit removal requirements are somewhat different for each application and loading rates can vary tremendously by site, event, and collection system configuration. In many HRT applications, the majority of the grit load is already being conveyed to the main headworks facilities and additional wet weather flows are essentially being "scalped" from the top of the collection system without a significant amount of grit to be handled by the HRT facilities. In other instances, the wet weather facilities should be designed to receive heavy grit loads. The HRT processes themselves are not particularly sensitive to grit; therefore, there are a wide variety of alternatives that should be considered, ranging from simply co-handling grit along with suspended solids (i.e., conveying grit with CES sludge or high-rate filtration backwash) to simple "rock boxes" or stilling wells to more complex grit removal basins (nonaerated, aerated, mechanically induced vortex, or hydraulic vortex).

3.6.2 Disinfection

Water quality studies often identify microbial pathogens as the principal parameter of concern for wet weather flows (WERF, 2009); therefore, effluent disinfection is of particular importance for most HRT facilities. Side-by-side piloting and full-scale operations demonstrated that the effluent from EHRT facilities had similar dose responses to hypochlorite, peracetic acid, or UV as the effluent from conventional secondary treatment processes for indicator bacteria (Crow and Coxon, 2011; Fitzpatrick, Andrews, Bahar, Jaworski, Tarallo, and Wagner, 2010; Nitz et al., 2004). Additional pilot and full-scale demonstration studies that include pathogenic bacteria, protozoa, and viruses and indicator organisms in EHRT applications have begun and further research in these areas is recommended (Black & Veatch, 2012; Gsellman and Dumbaugh, 2012). Refer to Chapter 15 for an additional discussion of effluent disinfection alternatives.

3.7 Biological Additions

In addition to Chapter 13 of this publication, refer to Chapter 13 and 14 of *Design of Municipal Wastewater Treatment Plants* (WEF et al., 2010) for further details of biological processes. Some of the CES technologies described in this chapter have been used for mixed liquor separation in suspended growth biological systems. The application of the ACTIFLO system in a temporary biocontact mode (BIOACTIFLO) was first piloted in the winter of 2004–2005 at the P Street WRRF in Fort Smith, Arkansas, (Sun et al., 2008), and has since been piloted in several other locations in the United States. The first two full-scale installations were constructed in 2012 at the Munster WRRF in St. Bernard Parish, Louisiana, and at the Wilson Creek WRRF in Texas (Sandino et al., 2011). A third installation is scheduled for commissioning at the Cox Creek WRRF in Anne Arundel County, Maryland (Perry, 2013). One of the main design considerations is that the mixed liquor alternative results in higher clarifier solids loading and slower settling floc, requiring approximately twice the settling area compared to microsand ballasted flocculation facilities without biocontact. In addition, the biocontact tank itself further increases the footprint requirements of the overall facility. In addition, ballast recovery is marginally less efficient in the mixed liquor alternative.

Magnetite ballasted flocculation has also been adapted to mixed liquor applications. Work by Woodard and Andryszak (2011) and Catlow and Woodard (2009) demonstrated that the BioMag process approximately tripled the solids loading capacity of conventional secondary clarifiers, significantly increasing capacity to handle wet weather flows.

In addition to the suspended growth alternatives mentioned previously, biofilm processes continue to be researched and developed. Piloting of the WWETCO FlexFilter in Springfield, Ohio, included periods of continuous operation, during which time biofilm growth was observed on the filter media, raising the possibility of a biofilm mechanism if operated continuously between events as an intermediate filter (Fitzpatrick et al., 2011). Variations of biologically active filters (BAFs) have also been developed and used at WRRFs, although their design is typically optimized for the treatment of dry weather flows. In fact, some of the upflow floating media filters mentioned earlier in this chapter were adapted from BAF technologies and optimized for wet weather flows.

Initial studies, including those referenced previously, collectively suggest that the biocontact addition tends to be capable of providing only marginal effluent quality improvements for peak wet weather flows compared to their core CES or filtration process at the expense of increasing facility footprint requirements and potentially making operations and process control more complicated. Further research is recommended to better understand the potential long-term advantages and disadvantages of biocontact addition, particularly for intermittent operations.

3.8 Dual Functionality

Besides wet weather flows, certain EHRT technologies may also be beneficial in the treatment of dry weather flows, thus providing more investment value. Many of the high-rate filtration alternatives have been demonstrated for tertiary filtration of secondary effluent (see Chapter 16 of *Design of Municipal Wastewater Treatment Plants* [WEF et al., 2010]). Some of the filter technologies have also been used with upstream chemical addition to further increase removal performance, particularly for phosphorus. Compressible media and cloth media filters have also been pilot tested as an intermediate step following primary clarifiers. Tchobanoglous (2011) suggested that primary effluent filtration should be further considered as a method to reduce the particulate and BOD loading to downstream biological processes, potentially increasing the energy efficiency of the overall WRRF, particularly if incineration or digester biogas is used for energy recovery. As illustrated in Figure 14.5, these dual-use concepts have also been demonstrated with CES

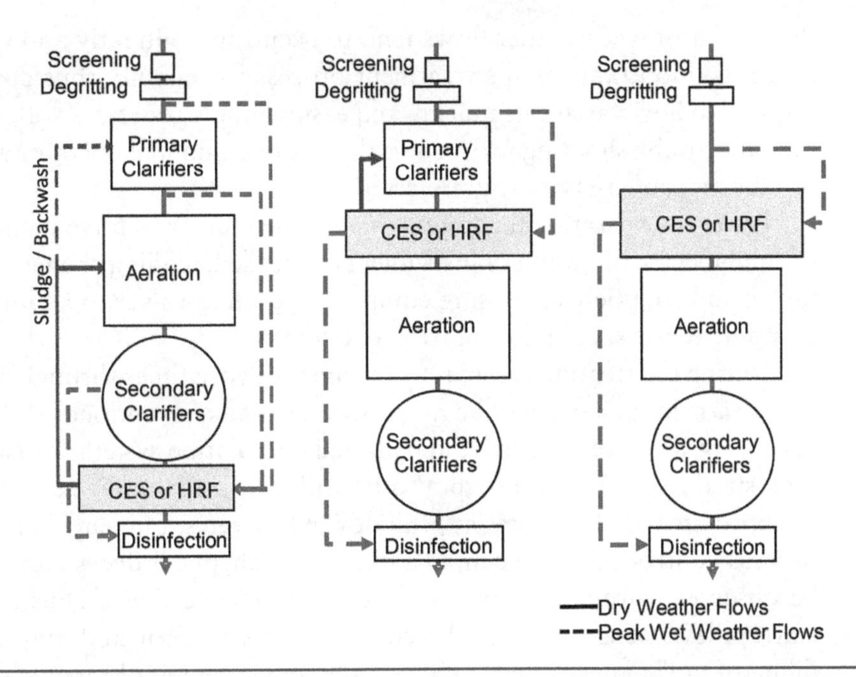

FIGURE 14.5 Chemically enhanced sedimentation and high-rate filtration alternatives also have the potential to be used in the treatment of dry weather flows. Pilot- and full-scale studies have demonstrated tertiary and enhanced primary treatment applications. Although not illustrated here, magnetite ballasted flocculation has also been integrated to the activated sludge process at some WRRFs to increase the loading capacity of secondary clarifiers.

alternatives, most notably for tertiary phosphorus removal and to decrease loadings to downstream aeration basins. Furthermore, the BioMag process has typically been implemented full scale as an integrated part of an activated sludge process and operated during both dry and wet weather conditions. Most of the BioMag installations have been primarily focused on increasing the capacity of secondary clarifiers.

Sedimentation tanks, CES, filters, and biofilters have also been used for the treatment of stormwater, an application that has similar technical requirements as peak wet weather flow treatment. A variety of different filtration media have been used and continue to be researched (Sileshi et al., 2011; Togawa et al., 2011). Examples of recent full-scale stormwater filtration systems include the Weracoba Creek facility in Columbus, Georgia; North Boeing Field facility in King County, Washington; Mar Vista Park facility in Santa Monica, California; and SABA Pfaffensteig, Gabelbach, and Wylerholz in Switzerland (Baumann, 2013; Boner et al., 2009; Magura and Shapiro, 2008; U.S. EPA Region 10, 2012).

4.0 OPERATIONAL CONSIDERATIONS

Because most wet weather flows tend to occur intermittently and vary rapidly, it is important for the treatment process to start up, shut down, and respond to flow variations quickly and easily. Simple processes with straightforward troubleshooting procedures that require minimal operator attention or adjustments are typically preferred.

Pre-event operational activities should include preventive maintenance of equipment, calibration of instruments as needed, refilling chemical storage tanks, and routinely exercising equipment (cycling valves, operating gates, bumping motors, etc.) during extended idle periods.

During the treatment event, hydraulic residence times through the high-rate treatment systems will be much shorter than conventional WRRF processes. For instance, a microsand ballasted flocculation system with a settling tank sized for 147 m/h (60 gpm/sq ft) will have a hydraulic residence time of approximately 8 minutes at peak design flowrates. Effluent changes from process control actions such as a change to chemical doses can typically be observed within approximately three residence times. During system startup, level sensors should be considered to monitor and automate the filling of tanks and sequence the starting and stopping of mixers, scrapers, pumps, and other equipment. Large variations in wet weather flowrates tend to require higher than normal turndown ratios, which may lead to more parallel chemical feed units and process trains than are typically seen in most WRRFs. Startup sequences, gate travel rates, and so on, should be adjusted so that filling an empty basin does not adversely affect the flow to other parallel basins already in service. Ballasted flocculation requires routine monitoring of the ballast concentration within the process. Extended event run times may require ballast makeup during an event, but, in many instances, ballast makeup can be completed either before or after the actual event.

Postevent requirements for system layup and preventive maintenance needs should also be considered. Each of the technologies described herein has slightly different layup and preventive maintenance recommendations. Equipment and system suppliers should be consulted to develop a postevent procedure for each application. In many instances, these procedures involve a sequence of basin and channel draining and flushing (see Chapter 7) unless the wet weather facilities are also designed to be used to treat dry weather flows (e.g., units used for phosphorus removal or tertiary clarification).

It should be noted that this publication focuses on wet weather flow treatment at an existing WRRF; however, some sewersheds may instead be best served by a satellite wet weather treatment facility. Many utilities do not permanently staff satellite facilities and operator response times, automation, and passive treatment processes may play a larger role in design decisions than with facilities that have operators on-site at all times. For further discussion of operational considerations for solids contact and ballasted flocculation facilities, refer to literature by Crow and Coxon (2011), Fitzpatrick et al. (2008), and Keller et al. (2005).

5.0 CASE STUDIES

Examples of EHRT technologies being used to treat peak wet weather flows at WRRFs include those described in the following subsections.

5.1 Chemically Enhanced Sedimentation with Sludge Recirculation and Solids Contact

The latest generations of this technology have been operating in municipal wet weather/primary treatment applications worldwide since the mid-1980s and in North America since the mid-1990s. One recent example is at the Bay View Water Reclamation Facility in Toledo, Ohio, which has a peak capacity of 1510 ML/d (400 mgd) and serves a collection system that is 30% combined with the following influent flow characteristics:

Average dry weather flowrate	170 ML/d (45 mgd)
Average daily flowrate	265 ML/d (70 mgd)
Maximum monthly average flowrate	492 ML/d (130 mgd)
Peak wet weather flowrate	1510 ML/d (400 mgd)

The auxiliary wet weather treatment facilities depicted in Figure 14.6 were commissioned in 2006 and include six parallel CES units, each with two coagulant rapid-mix chambers, two chambers for polymer mixing and solids contact/flocculation, and a common clarifier with tube settlers and sludge scrapers. The units use the proprietary DensaDeg process, with floc ballasting provided by sludge recirculation. A common-wall structure houses the CES units, a central pump and piping gallery, and polymer and ferric chloride chemical feed systems. Pretreatment is provided by six parallel vortex grit removal basins with a total peak capacity of 878 ML/d (232 mgd). Disinfection and reaeration of CES effluent is provided in a chlorine contact basin, followed by a 760-m (2500-ft) effluent conduit to the final effluent outfall channel.

One of either Units 1 and 3 is used for primary treatment of wet weather flows before activated sludge treatment. The other five units treat wet weather flows before on-site storage/equalization or effluent disinfection and discharge when the equalization basin is full. When influent flowrates approach 492 ML/d (130 mgd), the wet weather facilities are readied for operation. Once the flowrate reaches approximately 606 ML/d (160 mgd), the Bay View Pump Station is set to deliver flow to the wet weather

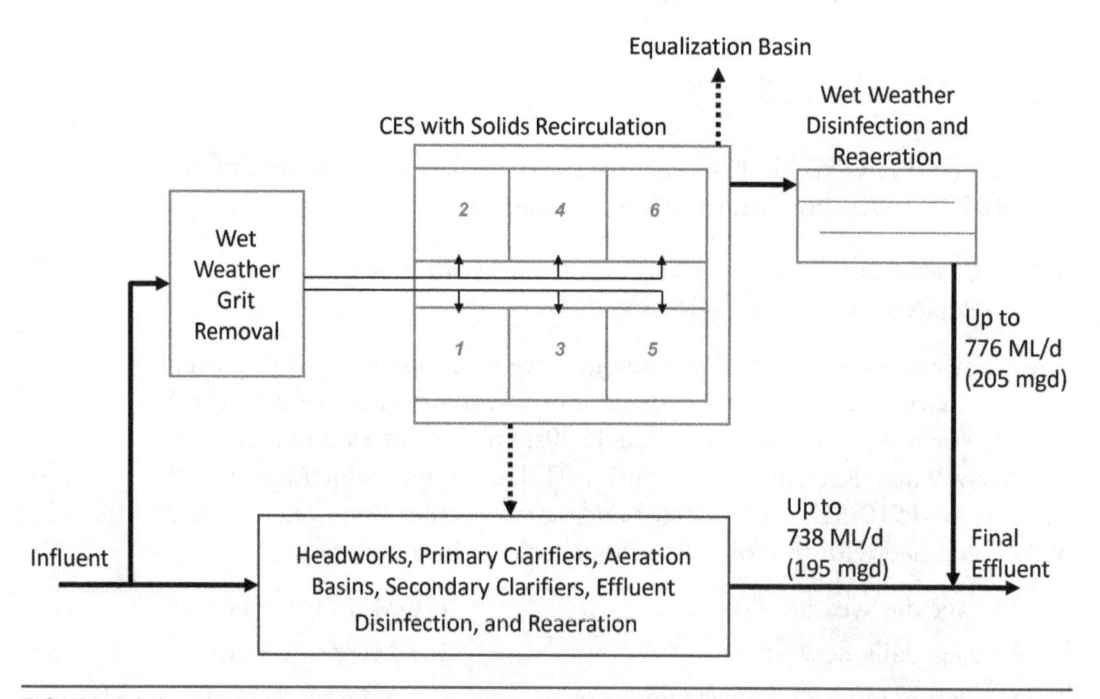

FIGURE 14.6 During peak wet weather flows, the Bay View Water Reclamation Facility in Toledo, Ohio, operates a solids contact CES system in parallel with its activated sludge facilities.

treatment facilities. Wet weather flow is also delivered to the facilities from the remote Windermere Pump Station, which is brought online when the Ten Mile Creek Interceptor becomes surcharged by rising wet weather flows. Flow to the wet weather treatment facilities is typically increased in approximately 95- to 114-ML/d (25- to 30-mgd) increments as additional pumps are started. Each time a new pump is started, an additional grit removal and CES unit are brought online. Results from a 2-year performance study (Black & Veatch, 2009) along with monitoring data from the subsequent 3 years of operation are summarized in Table 14.2. Operational lessons learned from this case study include the following:

- Influent turbidity measurements are erratic and unreliable for pacing chemical feeds. Effluent turbidity is less erratic, but does not correlate well to TSS.
- Ferric chloride dosages must be limited to prevent effluent pH violations. Influent pH is typically less than 7.0 during wet weather events.
- Ferric chloride to polymer dose ratios of 15:1 to 30:1 work best. Ferric chloride and polymer setpoints are typically set to 15 mg/L and 0.5 mg/L for primary treatment and 20 mg/L and 1.0 mg/L for discharge to effluent disinfection.
- When bringing a parallel unit into service, maintain the inlet gate at just 10% open until the level in all parallel units is equalized.
- Plan on foaming to occur in turbulent areas of effluent flow.

5.2 Chemically Enhanced Sedimentation with Microsand Ballasted Flocculation

ACTIFLO ballasted flocculation units have been operating in municipal wet weather treatment applications worldwide since the late 1990s and in North America since 2001. Pilot and full-scale case studies are well documented in literature, including the following:

- East Side Treatment Facility (Bremerton, Washington)—completed in 2001, this was one of the first full-scale facilities in the United States to treat wet weather flows using ballasted flocculation. It is a single-train facility with a peak flow capacity of 76 ML/d (20 mgd) and consists of a two-cell influent storage basin, side wiper hydraulic bar screen, coagulant addition, polymer and microsand flash mixing tank, flocculation tank, settling tank, and UV disinfection system. Although the East Side Treatment Facility is a satellite facility to Bremerton's Westside Wastewater Treatment Plant, its long-term performance is

TABLE 14.2 Toledo, Ohio, wet weather treatment facilities data summary.

Two-year performance study: April 2007 to March 2009					
		Treated and discharged		Treated and stored	
		Range of values	Average	Range of values	Average
Number of wet weather events		10		19	
Hours of operation		480		281	
Total volume treated, mil. gal		1001		556	
Influent flowrate, mgd (peak range, average)		77 to 167	50	42 to 165	48
Influent	TSS, mg/L	46 to 166	115	23 to 313	139
	CBOD, mg/L	26 to 57	38	34 to 109	64
	Particulate CBOD*, mg/L	17 to 45	27	14 to 84	45
	TKN, mg/L	4.6 to 8.5	6.2	6.7 to 13.2	9.8
	Total phosphorus, mg/L	0.7 to 1.6	1.0	0.8 to 1.9	1.3
Effluent	TSS, mg/L	11 to 43	28	9 to 75	27
	CBOD, mg/L	13 to 24	18	19 to 58	32
	Particulate CBOD*, mg/L	4 to 10	7	3 to 45	15
	TKN, mg/L	3.4 to 6.6	4.8	5.3 to 9.2	7.6
	Total phosphorus, mg/L	0.2 to 0.5	0.3	0.2 to 0.7	0.4
Operating data summary: April 2009 to March 2012					
Number of wet weather events		23		Not recorded	
Hours of operation		685			
Total volume treated, mil. gal		1317			
Influent flowrate, mgd (peak range; average)		28 to 236	41		
Influent	TSS, mg/L	40 to 380	154		
	CBOD, mg/L	20 to 130	50		
Effluent	TSS, mg/L	7 to 40	19		
	CBOD, mg/L	5 to 46	17		

*Carbonaceous biochemical oxygen demand (CBOD) concentration minus soluble CBOD concentration.

TKN = total Kjehldahl nitrogen.

representative of other wet weather applications and is summarized in Figure 14.7. Refer to literature by Crow and Coxon (2011) for further details.

- Lawrence, Kansas—wet weather flows exceeding the capacity of this WRRF's activated sludge trains are treated by 6-mm perforated plate screens, 2 × 76-ML/d (20-mgd) ACTIFLO trains, hypochlorite disinfection, and bisulfite dechlorination. Startup was in 2003 and case studies have been conducted by Keller et al. (2005) and Fitzpatrick et al. (2008).
- Village Creek Water Reclamation Facility (Fort Worth, Texas)—screened wet weather flows that exceed the capacity of the activated sludge trains can be treated through 2 × 208-ML/d (55-mgd) ACTIFLO trains whose effluent can either be stored in a 379-ML (100-mil. gal) storage basin or discharged through a chlorine contact basin. The ballasted flocculation upgrades were completed in 2005.

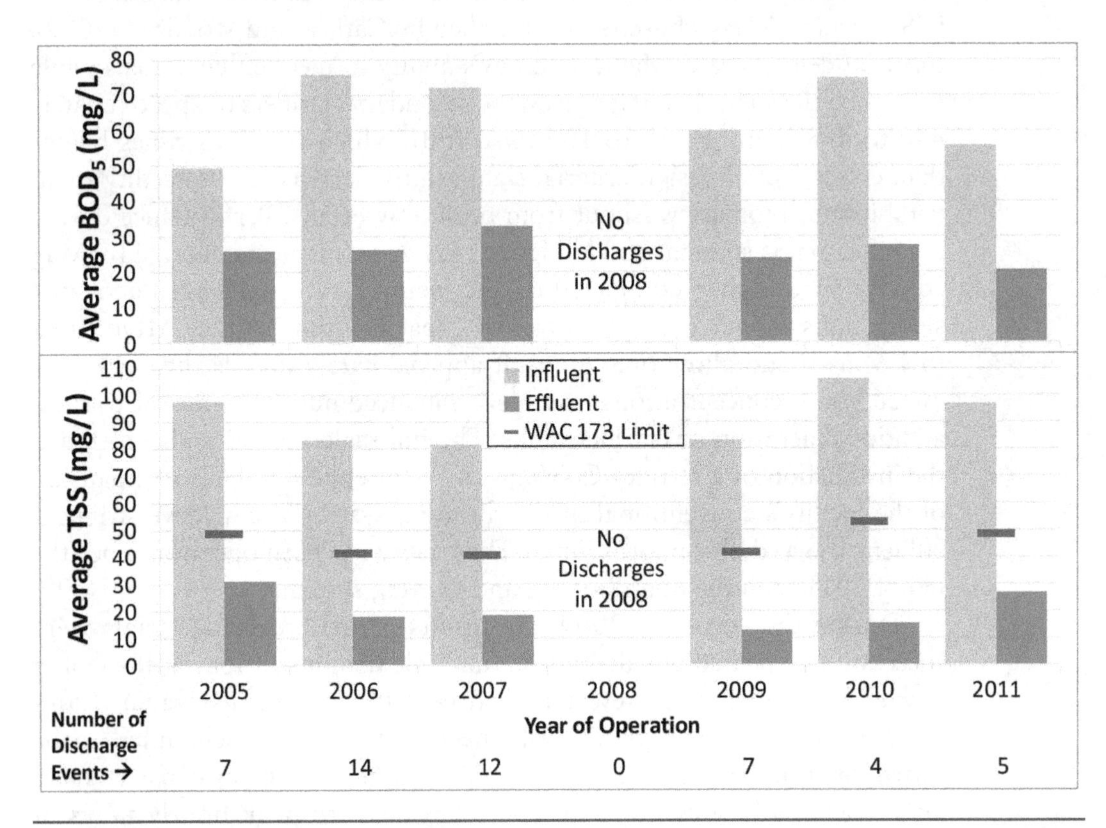

FIGURE 14.7 Long-term monitoring results from the microsand ballasted flocculation East Side Treatment Facility in Bremerton, Washington, demonstrated that the effluent quality consistently achieved applicable technology-based effluent limits (BOD_5 = 5-day biochemical oxygen demand; WAC 173 = Washington [State] Administrative Code, Title 173).

- Wilson Creek Regional Wastewater Treatment Plant (North Texas Municipal Water District)—commissioned in 2012, the single 121-ML/d (32-mgd) ACTIFLO train was designed as a dual-function facility, providing tertiary phosphorus removal of dry weather flows with the ability to be reconfigured to a biocontact mode (BIOACTIFLO) for treatment of wet weather flows. Refer to literature by Sandino et al. (2011) for further details.

5.3 Chemically Enhanced Sedimentation with Magnetite Ballasted Flocculation

Magnetite ballasted flocculation was first applied for tertiary phosphorus removal before being used to treat wet weather flows. Its first installation at a U.S. WRRF was in Concord, Massachusetts, in 2007. More recently, the technology has been used to ballast activated sludge to increase the solids loading and wet weather flow capacity of secondary clarifiers. A side-by-side full-scale trial of the BioMag technology was conducted for a 3-month period in Sturbridge, Massachusetts. As described by Catlow and Woodard (2009), this trial demonstrated the technology's ability to meet effluent goals while increasing the solids loading rate on the secondary clarifiers to approximately 440 to 488 $kg/m^2 \cdot d$ (90 to 100 lb/sq ft/d), which is 3 to 5 times higher than conventional design criteria, significantly decreasing the facility's vulnerability to biomass washout from peak flow events. Performance during the pilot trial is summarized in Figure 14.8. A permanent full-scale BioMag conversion was then completed on the facility's three packaged activated sludge units to increase overall facility capacity from 2.8 to 6 ML/d (0.75 to 1.6 mgd) and allow operations at approximately double their previous mixed liquor concentrations to achieve enhanced nutrient removal without additional aeration or clarifier tanks. The full-scale upgrades also included the installation of a tertiary CoMag system to replace a planned expansion of the facility's conventional media filtration system and achieve stringent effluent TSS and phosphorus limits. The facility has been operating since the fall of 2011 with the dual BioMag and CoMag system.

During the spring of 2009, the results of the Sturbridge trials were successfully repeated in side-by-side full-scale demonstrations at the Upper Gwynedd Township Wastewater Treatment Plant in Pennsylvania (Cambridge Water Technology, 2009). The retrofit there is scheduled to be operational in 2014 to increase the wet weather treatment capacity from an average daily flowrate of 11 ML/d (3 mgd) to peak hourly flows of approximately 45 ML/d (12 mgd), with effluent TSS objectives of less than 10 mg/L monthly average and less than 30 mg/L maximum daily average (Backman, 2013).

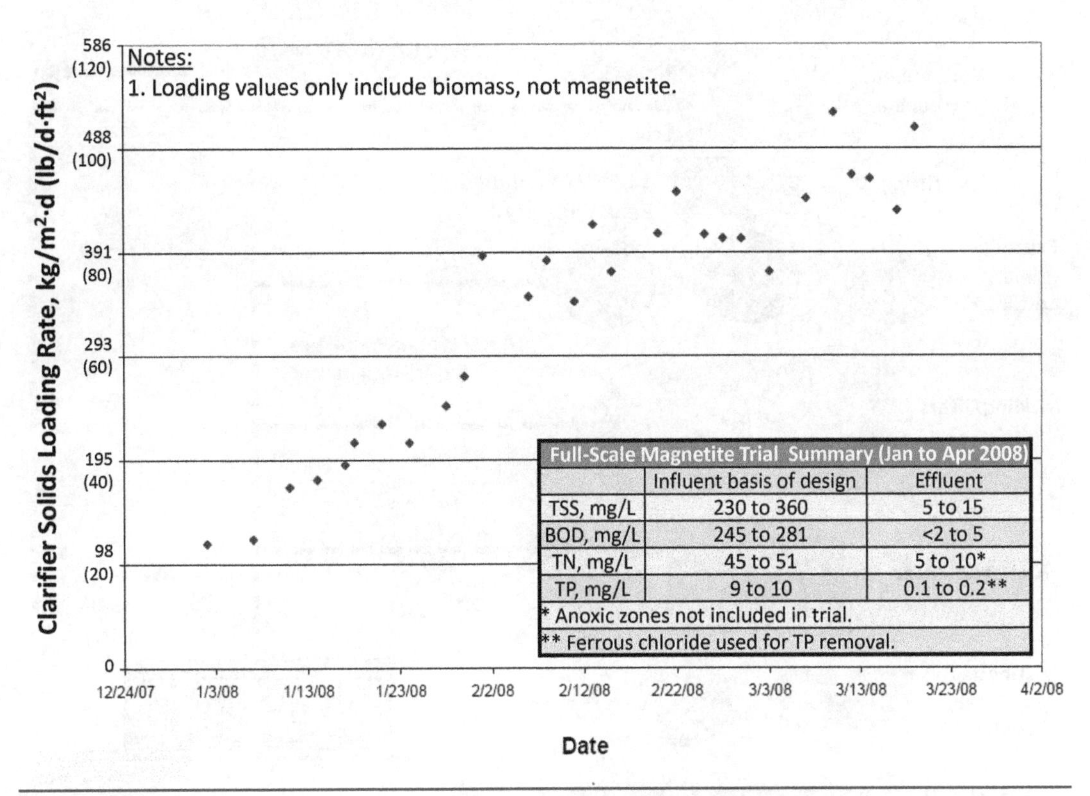

Full-Scale Magnetite Trial Summary (Jan to Apr 2008)		
	Influent basis of design	Effluent
TSS, mg/L	230 to 360	5 to 15
BOD, mg/L	245 to 281	<2 to 5
TN, mg/L	45 to 51	5 to 10*
TP, mg/L	9 to 10	0.1 to 0.2**
* Anoxic zones not included in trial.		
** Ferrous chloride used for TP removal.		

FIGURE 14.8 Full-scale side-by-side piloting in Sturbridge, Massachusetts, demonstrated the ability of magnetite ballasted flocculation (in activated sludge mixed liquor) to significantly increase the capacity of existing secondary clarifiers (adapted from Catlow and Woodard, 2009) (TN = total nitrogen; TP = total phosphorus).

5.4 Compressible Media Filtration

Following an 8-month pilot study with multiple wet weather events, the city of Springfield, Ohio, is constructing a compressible media filtration facility to increase the peak capacity of their WRRF (Fitzpatrick et al., 2011). As illustrated in Figure 14.9, the activated sludge facilities are rated for a peak flowrate of 130 ML/d (34 mgd) and the CMF facilities are designed for a peak capacity of 379 ML/d (100 mgd).

Combined sewers will flow through an influent control structure (ICS) equipped with four horizontally raked CSO screens along a side-exiting overflow weir. Flows to the WRRF are measured downstream of the main channel of the ICS. Two modulating gates maintain a flow setpoint to the activated sludge facilities. Flows that exceed the capacity of the activated sludge process overflow through the screens to the CMF facility. The screenings remain in the influent sewer and are conveyed to the existing headworks for removal. The ICS also has a depressed section across the bottom of the

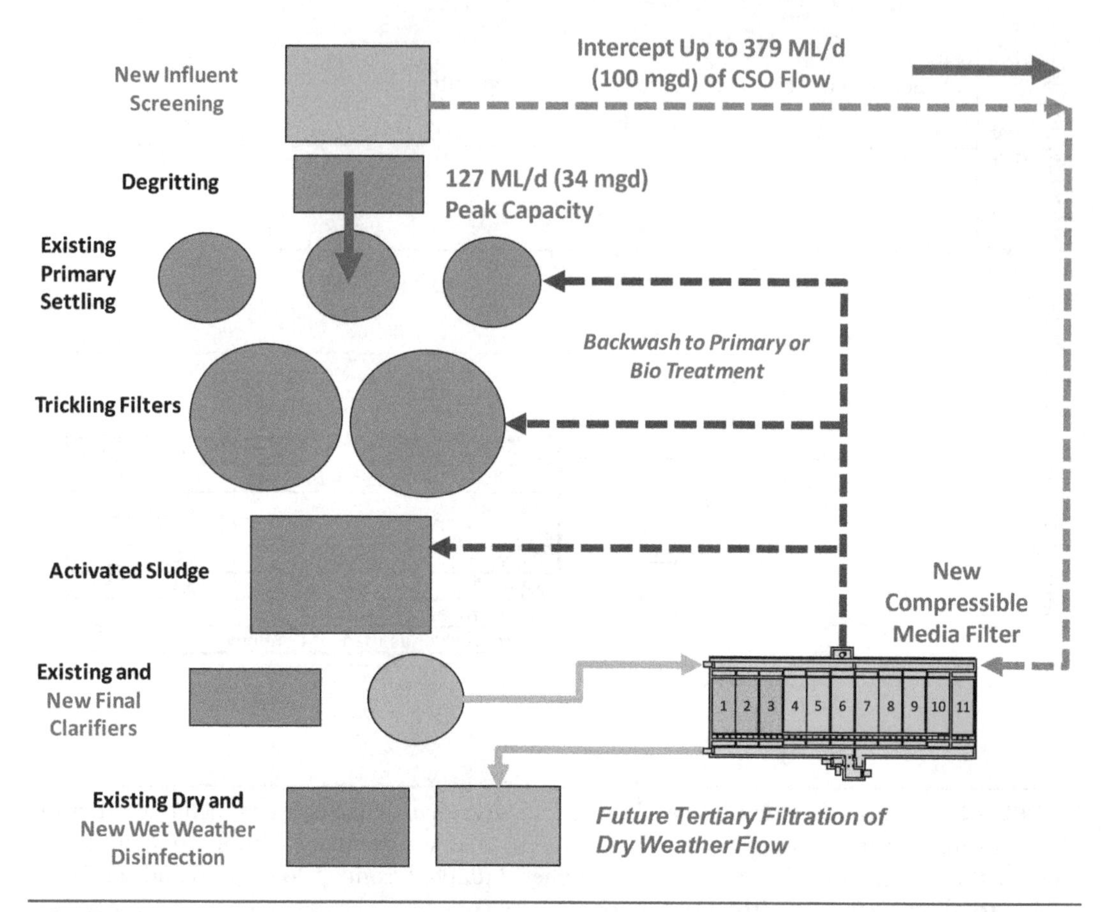

FIGURE 14.9 Following a successful 8-month pilot, the city of Springfield, Ohio, is constructing a compressible media filtration facility to increase the peak wet weather flow capacity of its WRRF.

main channel that collects rocks and other large settleable solids for periodic removal with a clamshell.

The new auxiliary wet weather treatment facility is a common-wall structure with 11 CMF cells, a chlorine contact basin, an effluent pumping station, and a backwash pumping station whose superstructure also houses backwash blowers and electrical gear above its wet well. A flow meter upstream of the wet weather treatment facility is used to determine the number of filter cells to place into operation. As the solids accumulate in the filter cells and increase the head over the filter, they will be sequenced for backwash. The backwash system is designed to allow up to two cells to be backwashed concurrently. Multistage centrifugal blowers provide the air scour for the backwash cycle. Flow from the filters continues to the chlorine contact basin for disinfection and is then pumped to the existing outfall pipe.

During typical operation, effluent from an adjacent filter cell is used for the backwash water source; however, effluent from the final clarifiers may also be used at the end of an event when filtered effluent may not be available. The final clarifier effluent connection can also be used in the future to allow a portion of the CMF cells to be used for tertiary filtration; however, this would require an effluent pumping station. The wet weather facility, including the chorine contact basin and effluent pumping station, is designed to drain to the backwash pumping station wet well at the end of a wet weather event.

6.0 REFERENCES

Backman, R. (2013) Siemens Water Technologies LLC. Personal communication.

Baker, M. N.; Taras, M. J. (1981) *The Quest for Pure Water—The History of Water Purification from the Earliest Records to the Twentieth Century, Volume 1 and 2;* American Water Works Association: Denver, Colorado.

Baumann, P. G. (2013) Aqua-Aerobic Systems, Inc. Personal communication.

Black & Veatch (2009) *Ballasted Flocculation Performance Testing 2-Year Final Report*, Report for the Bay View Wastewater Treatment Plant, Toledo, Ohio.

Black & Veatch (2012) *Pathogen Study 2011 Annual Report*, Report for the Toledo Waterways Initiative, Toledo, Ohio.

Boner, M.; Arnett, C.; Gurney, P.; Bowman, J. (2009) An Urban BMP for Bacteria and Aquatic Biology TMDLs. *Proceedings of the Georgia Water Resources Conference*; University of Georgia; Athens, Georgia.

Cambridge Water Technology (2009) *Upper Gwynedd Township Wastewater Treatment Plant Full Scale BioMag™ Trial*, Report for the Upper Gwynedd Township, Pennsylvania; Cambridge Water Technology: Cambridge, Massachusetts.

Catlow, I.; Woodard, S. (2009) Ballasted Biological Treatment Process Removes Nutrients and Doubles Plant Capacity. *Proceedings of Nutrient Removal (WEF Specialty Conference)*; Water Environment Federation: Alexandria, Virginia.

Crow, J.; Coxon, P. (2011) Ten Years of Operational Experience on HRC for Remote CSO System. *Proceedings of the 84th Annual Water Environment Federation Technical Exhibition and Conference* [CD-ROM]; Los Angeles, California, Oct 15–19; Water Environment Federation: Alexandria, Virginia.

Crow, J.; Smyth, J.; Bucher, B.; Sukapanotharam, P. (2012) Treating CSOs Using Chemically Enhanced Primary Treatment with and without Lamella Plates—How Well Does It Work? *Proceedings of the 85th Annual Water Environment Federation Technical Exhibition and Conference* [CD-ROM]; New Orleans, Louisiana, Sept 29–Oct 3; Water Environment Federation: Alexandria, Virginia.

Fitzpatrick, J.; Andrews, H.; Bahar, A.; Jaworski, L.; Tarallo, S.; Wagner, D. (2010) A Case for Keeping High-Rate Auxiliary Treatment Facilities in the Secondary Treatment Picture. *Proceedings of the 83rd Annual Water Environment Federation Technical Exhibition and Conference* [CD-ROM]; New Orleans, Louisiana, Oct 2–6; Water Environment Federation: Alexandria, Virginia.

Fitzpatrick, J.; Gilpin, D.; Kadava, A.; Kliewer, A.; Pekarek, S.; Schlaman, J.; Tarallo, S. (2010) Wet Weather Pilot Studies Demonstrate Effectiveness of High-Rate Filtration Technologies. *Proceedings of the 83rd Annual Water Environment Federation Technical Exhibition and Conference* [CD-ROM]; New Orleans, Louisiana, Oct 2–6; Water Environment Federation: Alexandria, Virginia.

Fitzpatrick, J.; Long, M.; Middlebrough, C.; Wagner, D. (2008) Meeting Secondary Effluent Standards at Peaking Factors of Five and Higher. *Proceedings of the 81st Annual Water Environment Federation Technical Exhibition and Conference* [CD-ROM]; Chicago, Illinois, Oct 18–22; Water Environment Federation: Alexandria, Virginia.

Fitzpatrick, J.; Wallis-Lage, C.; Bhandari, A.; Palomo, M. (2005) Ballasted Flocculation— Surfactants and Effluent Foaming Potential. *Proceedings of the 78th Annual Water Environment Federation Technical Exhibition and Conference* [CD-ROM]; Washington, D.C., Oct 29–Nov 2; Water Environment Federation: Alexandria, Virginia.

Fitzpatrick, J.; Weaver, T.; Boner, M.; Anderson, M.; O'Bryan, C.; Tarallo, S. (2011) Wet Weather Piloting toward the Largest Compressible Media Filter on the Planet. *Proceedings of the 84th Annual Water Environment Federation Technical Exhibition and Conference* [CD-ROM]; Los Angeles, Calfornia, Oct 15–19; Water Environment Federation: Alexandria, Virginia.

Gsellman, P. D.; Dumbaugh, T. (2012) Akron, Ohio Wet Weather History BIOACTIFLO™ Pilot Study Result. Presented at Five Cities Plus Conference, Columbus, Ohio.

Hattori, K. (2008) Solution for Sustainable Water. *Umwelttechnologie—Hessen trifft Japan*; Frankfurt. http://www.hessen-umwelttech.de/mm/Metawater_Hessen_trifft_Japan.pdf (accessed May 2014).

Hayashi, H. (2009) Current State and New Technologies for CSO Control in Japan. *Proceedings of the Japan–United States Joint Conference on Drinking Water Quality Management and Wastewater Control*; Las Vegas, Nevada.

Horie, N.; Shigemura H.; Hashimoto, T. (2011) CSO Control Policy and Countermeasures in Japan. *Japan–Korea Special Workshop on Impact Assessment and Control of Combined Sewer Overflow at the 4th IWA–ASPIRE Conference & Exhibition;* Tokyo, Japan. http://www.recwet.t.u-tokyo.ac.jp/e/symposium_e/4th_IWA-ASPIRE%20_Conference(E).html (accessed May 2014).

Kanaya, S. (2013) Metawater USA, Inc. Personal communication.

Keller, J.; Kobylinski, E. A.; Hunter, G. L.; Fitzpatrick, J. D. (2005) ACTIFLO®: A Year's Worth of Operating Experience from the Largest SSO System in the US. *Proceedings of the 78th Annual Water Environment Federation Technical Exhibition and Conference* [CD-ROM]; Washington, D.C., Oct 28–Nov 2; Water Environment Federation: Alexandria, Virginia.

Korean Environmental Protection Agency (2013) New Environmental Technology Verification No. 401—High Speed Upflow Filtration for Primary Treatment and CSO for BKT.

Liyanage, L. (2010) CSO Treatment at the City of Edmonton, Alberta, Canada. Wet Weather Treatment Options and Regulatory Strategies. *Proceedings of the 83rd Annual Water Environment Federation Technical Exhibition and Conference* [CD-ROM]; New Orleans, Louisiana, Oct 2–6; Water Environment Federation: Alexandria, Virginia.

Magura, L. M.; Shapiro, N. (2008) Meeting the City of Santa Monica's Urban Runoff Treatment Objectives. *Proceedings of the 81st Annual Water Environment Federation Technical Exhibition and Conference* [CD-ROM]; Chicago, Illinois, Oct 18–22; Water Environment Federation: Alexandria, Virginia.

Melcer, H.; Klein, A.; Land, G.; Butler, R.; Carter, P.; Ciolli, M.; Lilienthal, R. (2011) Revisiting a Wet Weather Option. *Water Environ. Technol.*, 23 (7), 56–61.

Min, J. H. (2013) BKT United. Personal communication.

Nitz, D.; Fitzpatrick, J.; Lyon, T. (2004) Wet Weather Treatment Pilot Testing: High Rate Clarification and UV Disinfection in a Cold Climate. *Proceedings of the 77th Annual Water Federation Technical Exhibition and Conference*; New Orleans, Louisiana, Oct 2–6; Water Environment Federation: Alexandria, Virginia.

Oij, E.; Banning, J.; Gore, J. (2010) The Use of Rapid Bioassessment to Assess the Success of Stormwater Treatment Technologies (Best Management Practices) in Urban Streams. In *Rapid Bioassessment of Stream Health*; Hughes D. L., Brossett, M. P., Gore, J. A., Olson, J. R., Eds.; CRC Press: Boca Raton, Florida; pp 169–184.

Perry, T. (2013) Kruger, Inc. Personal communication.

Sandino, J.; Covington, J.; Kraemer, J.; Boe, R.; McKnight, D.; Cole, B. (2011) First of its Kind Full-Scale Implementation of a Biological and Chemically Enhanced High-Rate Clarification Solution for the Treatment of Wet Weather Flows at a Municipal Wastewater Treatment Facility. *Proceedings of the 84th Annual Water Environment Federation Technical Exhibition and Conference* [CD-ROM]; Los Angeles, California, Oct 15–19; Water Environment Federation: Alexandria, Virginia.

Sileshi, R.; Pitt R.; Clark, S. (2011) Experimental Study on Particle Clogging of Biofilter Device in Urban Areas. *Proceedings of the 84th Annual Water Environment Federation Technical Exhibition and Conference* [CD-ROM]; Los Angeles, California, Oct 15–19; Water Environment Federation: Alexandria, Virginia.

Sun, J.; Townsend, R.; Parke, S.; Dillon, J. (2008) Biologically Enhanced HRC System Solving Peak Wet Weather Flow Challenges. *Proceedings of the 81st Annual Water Environment Federation Technical Exhibition and Conference* [CD-ROM]; Chicago, Illinois, Oct 18–22; Water Environment Federation: Alexandria, Virginia.

Tchobanoglous, G. (2011) Some Thoughts on the Future of Wastewater Treatment. *Proceedings of the Central States Water Environment Association 15th Annual Education Seminar;* Madison, Wisconsin.

Togawa, N.; Pitt, R.; Andoh, R.; Osei, K. (2011) New Research Focusing on Emerging Wet Weather Flow Management Strategies (UpFlow Filtration). *Proceedings of the 84th Annual Water Environment Federation Technical Exhibition and Conference* [CD-ROM]; Los Angeles, California, Oct 15–19; Water Environment Federation: Alexandria, Virginia.

U.S. Environmental Protection Agency (2013) *Emerging Technologies for Wastewater Treatment and In-Plant Wet Weather Management*; EPA-832/R-12-011; U.S. Environmental Protection Agency, Office of Wastewater Management: Washington, D.C.

U.S. Environmental Protection Agency Region 10 (2012) Slip 4 Early Action Area Web Site. http://yosemite.epa.gov/r10/cleanup.nsf/ldw/slip+4+early-+action+area (accessed July 2012).

U.S. Environmental Protection Agency (1978) *Field Manual for Performance, Evaluation and Troubleshooting at Municipal Wastewater Treatment*

Facilities; EPA-430/9-78-001; U.S. Environmental Protection Agency: Washington, D.C.

Water Environment Federation; American Society of Civil Engineers; Environmental & Water Resources Institute (2010) *Design of Municipal Wastewater Treatment Plants,* 5th ed.; WEF Manual of Practice No. 8; ASCE Manuals and Reports on Engineering Practice No. 76; Water Environment Federation: Alexandria, Virginia.

Water Environment Federation (2013) *Guide for Municipal Wet Weather Strategies;* Water Environment Federation: Alexandria, Virginia.

Water Environment Research Foundation (2002) *Best Practices for the Treatment of Wet Weather Wastewater Flows;* WERF 00-CTS-6; Water Environment Research Foundation: Alexandria, Virginia.

Water Environment Research Foundation (2003) *Peer Review: Wet Weather Demonstration Project in Columbus, Georgia;* Project 98-WWR-1P; Water Environment Research Foundation: Alexandria, Virginia.

Water Environment Research Foundation (2009) *Characterizing the Quality of Effluent and Other Contributory Sources during Peak Wet Weather Events,* Final Report; Project 03-CTS-12PP/PPa; Water Environment Research Foundation: Alexandria, Virginia.

Woodard, S.; Andryszak, R. (2011) Ballasted Biological Process Achieves Low Nitrogen and Phosphorus without Tertiary Filtration. *Proceedings of Nutrient Recovery and Management (WEF Specialty Conference)*; Washington, D.C.; Water Environment Federation: Alexandria, Virginia.

Yoon, Y. J.; Park, I. G.; Jung, M. K.; Rhu, D. H.; Eum, Y. J.; Min, J. H. (2012) Primary Treatment of Domestic Wastewater and Wet Weather Flow Using High Rate Up-Flow Filtration System with Floating Media in Mega City Seoul. *Proceedings of the 85th Annual Water Environment Federation Technical Exhibition and Conference* [CD-ROM]; New Orleans, Louisiana, Sept 29–Oct 3; Water Environment Federation: Alexandria, Virginia.

7.0 SUGGESTED READINGS

American Public Health Association; American Water Works Association; Water Environment Federation (2005) *Standard Methods for the Examination of Water and Wastewater*, 21st ed.; Eaton, A. D., Clesceri, L. S., Rice E. W., Greenburg, A. E., Eds.; American Public Health Association: Washington, D.C.

Fitzpatrick, J.; Bradley, P. J.; Duchene, C. R.; Gellner, J.; O'Bryan, C. R.; Ott, D.; Sandino, J.; Tabor, C. W.; Tarallo, S. (2012) Preparing for a Rainy Day—Overview of Treatment Technology Options for Wet Weather Flow Management. *Proceedings of the 85th Annual Water Environment Federation Technical Exhibition and Conference* [CD-ROM]; New Orleans, Louisiana, Sept 29–Oct 3; Water Environment Federation: Alexandria, Virginia.

15

Disinfection

Randall S. Booker, Jr., Ph.D., P.E.; Dempsey Ballou, P.E.; Annie Blissit, E.I.T.; Amanda Dobbs, P.E.; and Tim O'Brien, P.E.

1.0 INTRODUCTION

The purpose of wastewater disinfection is to achieve the permitted level of viable indicator organisms or target pathogens in the discharged effluent, thereby reducing opportunities for the spread of waterborne diseases and the potential risk to public health. Under wet weather conditions, significant challenges to achieving disinfection goals are presented, including high flows and atypical water quality characteristics. Few resources are available that focus specifically on the issues regarding disinfection under peak wet weather flows. In an effort to address this gap in knowledge, the goal of this chapter is to provide guidance regarding design and operational considerations for disinfection systems under wet weather conditions.

2.0 STANDARDS AND GUIDELINES

Design and operating standards related to wet weather treatment at water resource recovery facilities (WRRFs) are discussed in Chapter 2. Specified disinfection system requirements must typically be maintained during wet weather events; as such, developing site-specific design criteria that include wet weather data is critical. State standards affecting wet weather operations are limited to the specification of minimum contact times for chlorine-based disinfectants. Other relevant disinfection design criteria, such as chlorine dosage, UV light transmittance (UVT), and UV dose, are typically not addressed in relation to wet weather treatment and should be determined on a site-specific basis and validated with approving agencies and authorities during preliminary design (as should other important design criteria such as unit redundancy). Planning and design guidelines for disinfection systems with limited reference to wet weather design aspects are available and include the following reference documents:

- *Design of Municipal Wastewater Treatment Plants* (WEF et al., 2010);
- *White's Handbook of Chlorination and Alternative Disinfectants*, 5th Edition (Black & Veatch, 2010);
- *Recommended Standards for Wastewater Facilities,* also known as the 10 State Standards (GLUMBR, 2004);
- *Ultraviolet Disinfection Guidelines for Drinking Water and Water Reuse* (NWRI, 2003);
- *Wastewater Engineering: Treatment and Reuse,* 4th Edition (Metcalf & Eddy, 2003); and
- *Best Practices for the Treatment of Wet Weather Wastewater Flows* (WERF, 2002).

3.0 WET WEATHER DISINFECTION DESIGN CONSIDERATIONS

General wet weather treatment design considerations are presented in this section. Sections 5.0, 6.0, and 7.0 provide design information for specific disinfection technologies. The most significant effect from wet weather events on disinfection process selection is variability and range of influent wastewater characteristics. In addition, wet weather flows may have significant effects on upstream treatment systems, resulting in process upsets that affect influent to disinfection systems. A brief discussion regarding disinfection at combined sewer overflow (CSO) facilities is also presented in this section.

3.1 Basis of Design

The disinfection system design should be based on an adequate data collection campaign to characterize critical parameters under both dry weather and wet weather operating conditions. Guidance on the scope of data collection and evaluation to support design of wet weather treatment processes is included in Chapters 2 and 3. Appropriate disinfection process unit redundancy should also be considered. Disinfection systems may be designed with different levels of unit redundancy at different design conditions depending on regulatory requirements and owner/operator preferences. Preferred design practice is to meet peak flow conditions with the largest disinfection unit out of service. This approach may be more feasible for chlorine-based systems in which the unit of concern may be a chlorinator or chemical feed pump. With UV systems, this approach may be more cost- and space-prohibitive, in which case peak flow treatment may require all units in service. The ultimate determination should be made based on the actual wet weather peaking factor and definition of the critical unit of redundancy and reliability for the specific disinfection process. Pilot testing to support the design and operation of wet weather treatment systems may also be considered (Fitzpatrick et al., 2007).

3.2 Flow Variability

The range of flow variability resulting from wet weather events presents a significant design challenge with respect to disinfection systems. This potentially wide range of flows presents challenges to the cost-effective design of disinfection systems to meet performance goals, particularly with establishing sufficient contact time. Diversion of flow around the disinfection process is typically not an acceptable wet weather management strategy; therefore, the establishment of appropriate wet weather design flows for disinfection system sizing is critical. If improperly designed, disinfection systems may provide insufficient disinfectant dosage and contact time during wet weather events; they may also act as hydraulic bottlenecks, potentially resulting in surcharged upstream processes and overflows. Hydraulic effects and design flow development are discussed in Chapters 2 and 6.

To maintain disinfection performance across the design flow range, including wet weather events, disinfectant dose control is paced using flow, residual, and other process signals. Additional range of turndown may be achieved through the use of multiple process units and/or channels. Parallel processes may be used at some WRRFs to accommodate wet weather flows; in such cases, flow above a certain threshold may be diverted to a high-rate treatment process that is discharged back to the main treatment and disinfection process (ideally after storage and equalization). This mode of operation must be accounted for in the disinfection design basis through the data collection and evaluation phases of the project. If site conditions and cost considerations are favorable, on-site offline storage and equalization of wet weather flows (discussed in Chapter 10) may be an effective approach to managing flow variation and reducing effects to disinfection equipment sizing and operation.

3.3 Wastewater Characteristics

Influent wastewater characteristics vary during wet weather events. The initial phase of a wet weather event (the first flush) can create significant initial solids and organic loading to the WRRF, which can be associated with a higher indicator organism loading. After the first flush, sustained wet weather flows typically have a diluting effect; however, any reduction in effective loading to the WRRF will depend on the magnitude of flow and duration of the event. Site-specific studies are recommended to characterize wet weather flow events and provide guidance in designing and operating disinfection systems to achieve treatment goals under such conditions. Guidance on the wet weather treatment process design is presented in Chapter 2.

3.4 Upstream Process Effects

Wet weather events may affect upstream process performance and result in atypical influent wastewater characteristics to the disinfection system. Refer to Chapters 10 through 13 for a discussion of upstream process wet weather performance criteria. For example, increased ferric chloride addition as part of a wet weather management strategy in chemically enhanced primary treatment may result in a reduction of UVT and diminished disinfection performance.

Coupled with high wet weather flows, the first flush loading may result in upstream process effects or upsets that yield higher indicator organism concentrations to the disinfection process. In addition, first flush influent loading and associated upstream process effects may result in increases in total suspended solids (TSS), turbidity, organics, and other constituents that increase disinfectant demand and reduce disinfection system performance (Ting, 2012). For example, excessive overflow rates in secondary clarifiers, resulting from wet weather hydraulic loading, may result in increased effluent solids that reduce the UVT or increased chlorine demand.

The potential for partial or complete diversion of flows around one or more upstream processes must also be considered in the design of disinfection systems (U.S. EPA, 2005). A recent study evaluated the public health risk effects of the practice of wet weather blending of primary-treated and secondary effluents before disinfection. The study also includes guidance on the assessment of blending effects at WRRFs (WERF, 2009). Blending practices vary by state, and effects on disinfection system design and operation should be considered if the practice is permitted at a given WRRF. Guidelines on blending practices are pending final decisions by U.S. Environmental Protection Agency (U.S. EPA).

3.5 Combined Sewer Overflow Disinfection

Treatment operations at a CSO facility are intermittent and occur only when significant wet weather events occur, whereas WRRFs must maintain continuous treatment operations and be designed and operated to accommodate wet weather events. Disinfection processes at both types of facilities must be designed to achieve performance requirements under wet weather flows and characteristics; however, the range of operation or turndown required for disinfection systems at WRRFs may be greater (and thus pose greater design and operating challenges) than those at CSO treatment facilities.

The design and operation of disinfection systems at CSO facilities is also affected by the remote location and infrequent use of treatment systems. Infrequent use may affect aspects of monitoring, control, and long-term chemical storage. Combined sewer overflow treatment facilities are typically designed for high-rate physical treatment (sedimentation, flotation, filtration); these processes affect disinfection process selection. While the focus of this chapter is on WRRFs, disinfection process selection for CSO facilities is similar in approach. Aspects of disinfection process selection for CSO facilities are addressed in several referenced publications, including the U.S. EPA Combined Sewer Overflow Technology Fact Sheet on Alternative Disinfection Methods (1999a).

4.0 DISINFECTION PROCESSES

Chlorine-based disinfectants (primarily in the forms of gaseous chlorine and sodium hypochlorite) are the predominant wastewater disinfection methods in the United States. A 2008 Water Environment Research Foundation report indicated that approximately 75% of significant WRRFs surveyed used some form of chlorine for disinfection, while approximately 21% used UV for disinfection (WERF, 2008). The report also noted that the use of gaseous chlorine at WRRFs had declined significantly over the previous two decades, with a corresponding increase in the use of sodium hypochlorite and UV. Other, less commonly used disinfectants include ozone and chlorine dioxide. These disinfection processes, particularly aspects of their design and operation relevant to wet weather treatment, are described in the following sections.

5.0 CHLORINATION

Chlorination is the most common disinfection process used at WRRFs. Chlorine-based disinfectants consist of the following:

- Elemental chlorine (Cl_2) stored as a liquid delivered in rail car tanks, 1-ton cylinders, or 150 lb cylinders and typically fed as gaseous chlorine;
- Liquid sodium hypochlorite (NaOCl) delivered in bulk or totes typically in a nominal 12.5 to 15% concentration; and
- On-site generation (OSG) of sodium hypochlorite from the electrolysis of salt and water.

5.1 Chlorination System Design

The chlorine disinfection system must be designed to work under all operational conditions, from low flows during early morning hours to peak wet weather flows during severe wet weather events. The challenge lies in optimally (and cost-effectively) sizing contact tanks, bulk chemical storage, chemical feed equipment, mixing systems, and control systems such that the wastewater is exposed to an adequate concentration of disinfectant for the proper amount of time over a wide range of flows.

The variable influent water quality of wet weather flows results in design and operational challenges. For example, wastewater pH affects the equilibrium concentrations of hypochlorous acid and hypochlorite ions. A high pH results in the hypochlorite ion being the predominant form present, which leads to a reduction in the hypochlorous acid as a disinfecting agent. The variations in wet weather pH are site-specific, and sampling is recommended to determine specific chlorine demand under various flow regimes.

Ammonia and other organic compounds also affect chlorine disinfection. An increased ammonia concentration leads to the formation of chloramines, thereby reducing the disinfectant's ability to kill or inactivate microbes (Boczek et al., 2008; U.S. EPA, 1999b). Like pH, wet weather ammonia is site-specific and could vary based on the efficacy of upstream unit processes (see also Chapters 11 through 14).

Wet weather flows are often associated with variable TSS concentrations depending on the duration and phase of the event and site-specific characteristics. A high solids content, typically found during the first flush, results in the shielding of microorganisms and increases the chlorine requirement (U.S. EPA, 1999b). Larger particles associated with a first flush condition can shield microorganisms from contact with chlorine and can also exert a chlorine demand (WERF, 2008). Additionally, during wastewater blending, particulate matter within primary effluent can reduce the effectiveness of disinfection (shielding effect). The shielding effect is more significant for virus particles than for bacteria (Boczek et al., 2008).

As stated previously, wet weather flows may affect secondary clarifier performance, resulting in higher overflow and solids loading rates and diminished settling performance and increasing the TSS in the secondary effluent, potentially resulting in reduced disinfection performance. The effectiveness of a chlorine-based disinfection process during a wet weather event is dependent on secondary clarifier performance and other upstream treatment processes.

Because wet weather influent water quality is highly variable and site-specific, wet weather chlorine demand should be established based on the results of bench testing. Bench tests should be performed during various times throughout wet weather events, particularly during the first flush.

5.1.1 Feed Equipment Selection

Chlorine demand may vary significantly during a wet weather event; often, the chemical is overdosed for chlorination and dechlorination (if required) to ensure a proper level of treatment. However, overdosing chlorine can lead to the formation of disinfection byproducts (DBPs); the WRRF must achieve a balance between adequately dosing for pathogen inactivation while limiting DBP formation.

Chlorine feed equipment is discussed in *Design of Municipal Wastewater Treatment Plants* (WEF et al., 2010). To properly size a chlorination system for wet weather events, the required feed rates/turndown must be determined for both dry weather conditions and wet weather conditions, including peak flow, particularly during first flush. Resulting from potentially greater chlorine demand and reduced available contact time, higher chlorine dose concentrations may be required under wet weather conditions to ensure microorganism kill under rapidly varying influent flow and quality. Given this higher dose requirement during peak flow, the turndown range requirement may be significantly wider for wet weather systems than for dry weather systems. Chlorinators, vaporizers, and sodium hypochlorite feed pumps must be sized to account for peak dose considerations, and redundancy must also be provided. Water resource recovery facilities may have multiple redundant chlorinators and vaporizers with full online capacity to achieve wet weather dose requirements, particularly during the first flush.

5.1.2 Mixing Equipment

It is important to ensure that the chlorine and water are completely mixed before entering the contact chamber. If the chlorine is not uniformly mixed, then even high doses will not be effective because the chlorine is not coming into contact with the entire flow. For optimum performance, a chlorine disinfection system should simulate plug flow characteristics and be highly turbulent for complete initial mixing in less than 1 second. The goal of proper mixing is to enhance disinfection by initiating a reaction between the free chlorine in the chlorine solution stream with the ammonia nitrogen. This prevents prolonged chlorine concentrations from existing and forming other chlorinated compounds (U.S. EPA, 1999b).

Studies have shown that high-rate disinfection processes that include increased mixing intensity and increased disinfection concentration are successful (U.S. EPA, 2007). High-intensity mixing using induction mixers for chlorine solutions, sodium hypochlorite, and dechlorination chemical injection has shown good performance in meeting effluent standards at high flowrates with short detention times (U.S. EPA, 2002). This allows WRRFs to provide both dry and wet weather disinfection in conventional chlorine contact basins (WEF, 2013).

5.1.3 Contact Tank

The chlorine disinfection system must be designed to work under all operational conditions, from low flows during early morning hours to peak wet weather flows. The challenge lies in optimally and cost-effectively sizing contact tanks such that the water is exposed to the proper concentration of disinfectant for the proper amount of time over a wide range of flows.

General guidelines for minimum chlorine contact time can be found in the "Ten State Standards" (GLUMRB, 2004) and *Design of Municipal Wastewater Treatment Plants* (WEF et al., 2010). State regulatory agencies may have specific design contact time and residual requirements. The primary consideration for wet weather contact tank design is to verify that the actual modal contact time is adequate for peak flow operation. Modal contact time may be assessed using computational methods and demonstrated using a tracer test. Additional chlorine contact time could be achieved during peak wet weather flows by adding chlorine before flow equalization or the gravity settling stage (WEF, 2013) or by using multiple parallel reactors (WEF et al., 2010).

Computational fluid dynamics models are increasingly being used to optimize disinfection reactor design. Computational fluid dynamics modeling can provide detailed hydraulic analyses under various flow regimes and identify hydraulic inefficiencies of the conventionally designed system, which will be exaggerated under wet weather peak flow conditions.

5.2 Chlorination Operation and Control

5.2.1 Analytical/Residual Instrumentation

Wet weather flows can have a significant effect on the accurate measurement of chlorine residuals, which is a key characteristic in controlling the disinfection process. A discussion regarding chlorine residual analyzers and oxidation–reduction potential analyzers can be found in *Disinfection of Wastewater Effluent—Comparison of Alternative Technologies* (WERF, 2008) and "Hypochlorite Conversion in Portland, ME Compliance, Wet Weather Disinfection, Improved Automation, and Success" (Birkel et al., 2007). Chlorine residual analyzers are sensitive to water quality variability, relatively high TSS, variability in color, and other water characteristics. The effects of wet weather flows can be mitigated by ensuring the analyzers are properly maintained and frequently cleaned to remove solids buildup.

5.2.2 Feed Control Strategies

With proper sizing and design of the chemical feed system and proper operation of the control system, chlorine doses should be readily adjustable to fluctuations in flowrate and contact tank influent water quality (WERF, 2008). There are several ways to control the feed rate of chlorine gas or sodium hypochlorite solutions, such as manual control, flow-proportional control, residual control, and compound-loop control. These various feed control strategies are presented in *Design of Municipal Wastewater Treatment Plants* (WEF et al., 2010).

Manual control is best used when the flowrate and demand are predictably consistent and is, therefore, not recommended for infrequent and highly variable wet weather flows. In flow-proportional control, the chlorine feed rate is determined based on the flowrate; however, this method is not always appropriate because chlorine demand can vary independent of flowrate depending on the wastewater characteristics. Residual control involves varying the chlorine feed rate based on the deviation of concentration from a setpoint on a controller. Residual control systems do not react well to large variations in flowrate over short periods of time and, therefore, are not recommended for wet weather flow conditions.

Compound-loop control uses both flow and residual input to regulate the chlorine feed. Compound-loop control is flow-paced, but uses the residual to trim the feed. Because this type of system provides more accurate control and can adjust quickly to changes in flowrate, compound-loop systems are recommended for treating wet weather flows at WRRFs. However, more training and an increase in the skill level of operating personnel is required for this more complicated system.

5.2.3 Storage

"Ten State Standards" (GLUMRB, 2004) and *Design of Municipal Wastewater Treatment Plants* (WEF et al., 2010) provide design and operational guidance on general chlorine storage. Wet weather treatment results in additional storage challenges at WRRFs. There are numerous safety considerations and storage capacity concerns associated with liquid chlorine, whether it is stored in cylinders, ton containers, or tank cars. A WRRF must consider how to safely and optimally store an appropriate volume of liquid chlorine to treat wet weather flows.

Sodium hypochlorite is typically delivered at a 12.5% nominal concentration; however, new technologies can provide higher concentrations up to 16%. Water resource recovery facilities must ensure that a sufficient volume is present on-site to treat extended wet weather events while also considering degradation. Because the solution strength degrades slowly over time, bulk quantities of sodium hypochlorite are typically not stored for periods longer than 60 days (WEF et al., 2010). It should be noted that the degradation potential is accelerated for higher concentration solutions. Because increased temperatures accelerate the degradation, sodium hypochlorite should be stored in the shade and its exposure to direct sunlight should be limited. Water resource recovery facilities can also minimize sodium hypochlorite degradation by using multiple storage units and shorter residence times, receiving more frequent bulk deliveries, and instituting a standard operating procedure (SOP) to rotate storage tanks and maintain optimal concentrations. Water resource recovery facilities can also dilute the sodium hypochlorite on-site to 6% or less to reduce the effects of degradation (Black & Veatch, 2010). However, significant care must be taken when diluting sodium hypochlorite solutions to avoid introducing contaminants that can catalyze the hypochlorite degradation.

5.3 On-Site Generation of Sodium Hypochlorite

General guidance in the design and operation of OSG systems and potential advantages and disadvantages of OSG relative to bulk storage systems are available (Black & Veatch, 2010; Metcalf & Eddy, 2003; WEF et al., 2010). A key consideration for OSG systems with regard to wet weather disinfection is the capacity of the sodium hypochlorite generation system and associated storage tanks. Typical OSG systems produce 0.8% sodium hypochlorite solutions and the OSG system and associated storage capacity must be sized to accommodate potentially increased and extended wet weather feed requirements at this solution strength. Newer OSG technology can generate 12.5% or higher concentrations. At 12.5% concentration, storage and feed considerations are identical to bulk sodium hypochlorite disinfection system designs.

It should also be noted that OSG systems require backup power, and wet weather events are often coupled with power outages. Backup power requirements should be based on typical power outage duration and the available storage of sodium hypochlorite generated before a wet weather event. At a minimum, backup power is required for the storage and feed pump systems.

5.4 Chlorination Technology Selection

The selection of a particular chlorination process should consider operation strategies and challenges inherent with wet weather flows, particularly with regard to sufficient mixing, minimum contact time, and efficacy of upstream unit process operations. Each technology has various advantages and disadvantages related to wet weather treatment; those advantages and disadvantages are summarized in Table 15.1.

5.5 Dechlorination

Dechlorination of WRRF effluent may be necessary to reduce toxicity resulting from the chlorine residual. Dechlorination system selection, design, and operational guidance is covered extensively in referenced material (Black & Veatch, 2010; WEF et al., 2010). During wet weather events, chlorine is

TABLE 15.1 Chlorination technologies and advantages/disadvantages related to wet weather disinfection.

Technology	Wet weather treatment advantages	Wet weather treatment disadvantages
Chlorine	Proven technology Flexible turnup/turndown ratios on chlorinators and vaporizers to meet peak disinfection demands under first flush and sustained wet weather flows Minimal bulk chemical storage footprint requirements	Significant safety hazards Complex operation
Sodium hypochlorite	Simple system Safety concerns minimized Peak doses can be achieved with larger pumps	Storage and shelf-life considerations Pumping/storage capacity requirements (12.5% concentration may require significant storage capacity to meet peak disinfection demands)
On-site generation	Minimizes transport of hazardous chemicals by generating chemicals on-site Provisions for delivery of bulk 12.5% sodium hypochlorite during generation equipment outages	Complex system Puming/storage capacity requirements (dilute vs concentrated OSG) Power requirements (higher risk of power outages and need for emergency power during wet weather events)

often overdosed to achieve microorganism inactivation during the rapidly varying influent flow and water quality conditions. Therefore, WRRFs that dechlorinate must have provisions to apply higher than typical doses of dechlorinating chemical to ensure an acceptable chlorine residual in the effluent stream. Reaeration of the effluent must also be considered when dechlorinating to meet the WRRF's permitted dissolved oxygen limit in the effluent.

6.0 ULTRAVIOLET LIGHT

Although the prevalence of chlorine disinfection systems has declined in the past two decades, the implementation of UV disinfection processes has grown tremendously, and accounting for the disinfection technology in more than 20% of all significant publicly owned treatment works in North America (WEF et al., 2010). Further, *Design of Municipal Wastewater Treatment Plants* (WEF et al., 2010) provides a detailed review of UV disinfection technology options, configurations, and system design. This section will focus on the design of a UV disinfection process as it specifically relates to wet weather flows.

6.1 Design Basis

6.1.1 Data Collection

Wet weather presents a challenge for UV system design because of the presence of a first flush, typically characterized by a low UVT and high TSS, followed by higher UVT waters as the wet weather event progresses and becomes predominantly stormwater. During wet weather events, UVT can vary from less than 20% during the first flush to more than 65% later in the same event (Muller and Lem, 2011). Water quality influent to a UV system will also be heavily affected by secondary clarifier performance during a wet weather event (Hunter, 2012). Higher overflow and solids loading rates and potentially diminished settling performance may result in excess TSS in the secondary effluent, which may reduce the UVT and enhance particle shielding and UV light scatter (Bell et al., 2011).

Sampling during wet weather events is critical in the design of a UV system because the variation of water quality during wet weather events is both site- and event-specific. Sampling of UVT and TSS is recommended every 10 to 15 minutes during a wet weather event (Muller and Petri, 2013; NWRI, 2003). Alternatively, automatic samplers can be installed to assist the operations staff with sample capture during wet weather events.

6.1.2 Dose Determination

Performing collimated beam studies throughout a wet weather event provides information for selecting the appropriate UV dose for design. A collimated beam test uses a bench-scale apparatus, which applies various known UV doses to the site's water samples to accurately determine UV dose response for that particulate water. Testing at least once early in the event and once at the end of the event are the minimum recommended tests. Ideally, collimated beam testing would occur at an early stage, during, and toward the end of a wet weather event. Conducting an additional collimated beam study during dry weather is also recommended (NWRI, 2003). After the tests have been completed, the appropriate dose may be selected from the dose response curves for each wet weather event phase or a single dose selected may be from a series of curves. One source for a testing protocol is U.S. EPA's Environmental Technology Verification program publication, *Generic Verification Protocol for High-Rate, Wet-Weather Flow Disinfection Applications* (Scheible and McGrath, 2000). Testing should include collimated beam testing, pilot performance tests, and fouling/wiper efficiency tests (Chandler et al., 2004). Additional guidance on UV system validation may be found in National Water Research Institute guidelines (NWRI, 2003).

If collimated beam tests are not feasible, WRRFs may reference a manufacturer-owned database containing data collected over a variety of water qualities, sources, locations, and types of upstream treatment processes to find comparable WRRFs to determine an appropriate dose (Muller and Lem, 2011). If wet weather flows go through the same upstream processes as dry weather flows, extrapolating doses is more reliable; however, direct scaling by multiplying the number of lamps associated with dry weather treatment by the wet weather flow factor, although safe, may oversize the UV system (Muller and Petri, 2013). Computational fluid dynamics modeling is used by some manufacturers to limit the flow that receives less UV dose as a result of short-circuiting or exposed lamps (Muller and Lem, 2011).

6.1.3 Reactor Hydraulics

Dispersion, turbulence, effective volume, residence time distribution, and flowrate should be considered in the hydraulic design of a UV system (Chandler et al., 2004). The design should optimize the effective water layer between the UV lamps and minimize the potential short-circuiting area (Muller and Lem, 2011).

A low first flush UVT requires a smaller effective layer; the wastewater present during extended wet weather events has a higher UVT and can be treated effectively with a larger effective layer. Treatment of the lower UVT waters can be achieved through more powerful lamps, narrower spacing, and the use of hydraulic devices to direct flows toward the lamps. However, these options come at a cost, whether it is higher energy consumption or increased headloss that must be considered (Muller and Lem, 2011).

Lamps should be spaced evenly apart and optimized to balance the tradeoff between headloss generated and mixing induced by closer spacing (Muller and Lem, 2011). Mixing to help evenly distribute the flow may be provided or even required for low UVTs as a part of the manufacturer's UV system (Muller and Petri, 2013). Increased headloss may lead to water level increases in open channel UV reactors, which may cause short-circuiting or exposure of downstream lamps to air (Muller and Lem, 2011).

6.2 Ultraviolet System Selection

6.2.1 Lamp/System Type

Ultraviolet systems vary by lamp technologies, reactor configurations, lamp orientation, cleaning mechanisms, and ballast design. The most common source of UV is the mercury vapor, electric discharge lamp with low- or medium-pressure configurations. Low-pressure systems are available with low-intensity and high-intensity lamps. Both use monochromatic germicidal light, but the high-intensity lamps have an output that is 1.5 up to 10 times per centimeter that of the low-intensity lamps. Medium pressure differs in that it uses polychromatic light and has an output that is magnitudes higher than low-pressure lamps. Although only 15% of the energy is converted to germicidal light, the higher pressure disinfection systems can be more effective at high TSS and low UVT levels (Chandler et al., 2004).

6.2.2 Lamp and Reactor Configuration

Open channel UV systems are more commonly used in wastewater applications than inline reactor systems; however, a closed vessel system provides a better level of control, which would benefit a system experiencing wet weather (Chandler et al., 2004). "Noncontact" UV systems in which water flows through the reactor in tubes made of a UV-transmissive material are also available; headloss restrictions at higher flow applications are important considerations for these systems (Black & Veatch, 2010). Multiple channels may be used to accommodate higher flows during wet weather. The ability of a WRRF to accommodate multiple channels and systems in parallel vs units in series is a function of hydraulic profile constraints, including available headloss, system space, and layout requirements. Using multiple channels in parallel offers the added benefit of bypassing, which enables one channel to be isolated and removed from service for maintenance. Flow should be evenly distributed to the channels to maintain an acceptable level of disinfection. This can be achieved through the use of influent channels, stilling plates, weirs, and gates.

6.3 Ultraviolet System Control

Manual control of UV systems is not preferred during wet weather events because of the variability of influent flow and water quality and the effects on upstream process performance. Automatic flow- and UVT-paced system control offers the highest flexibility for responding to quickly developing peak flow events. However, manual control may be required to override nonresponsive automatic controls or failed instrumentation during such events (Wood et al., 2005).

Flow pacing allows for alteration of light intensity and the number of lamps in use as a function of flow, as monitored by a weir or flume. U.S. EPA recommends quarterly calibration of flow meter instrumentation if flow pacing is used (Wood et al., 2005). However, flow pacing alone does not account for variations in UVT; as such, this may lead to underdosing or overdosing depending on the degree of variability in influent characteristics.

Dose pacing is considered the most sophisticated form of control for UV systems. The dosage is based on flow, lamp output, and water conditions. This method is used for low-pressure/high-output and medium-pressure systems. Because the dose is a function of water conditions, dose pacing is recommended for wet weather applications to account for the variable UVT throughout wet weather events. Flow monitoring and intensity and transmittance instruments are required for implementation. Power savings and lower maintenance costs are advantages of this control method because only the proper dose required is used to treat the wastewater. However, depending on site-specific equipment, staffing, and operational considerations as well as historical wet weather event frequency and duration, there may be advantages in simply operating the UV system at maximum output during wet weather events. The potential advantages (reduced operational and control complexity, reduced risk of insufficient dosing) and disadvantages (power cost, effect on equipment life, reliance on wet weather SOPs to override automatic dose control) should be assessed before implementing this approach.

7.0 ULTRAVIOLET DISINFECTION AND CHLORINATION FOR WET WEATHER APPLICATIONS

Flow and water quality variability during wet weather events will affect disinfectant demand and dose requirements for both UV and chlorine-based disinfection systems. Higher disinfectant demand in chlorine-based systems may often be accommodated by sufficient turndown range in the chlorination feed system, and reduced available contact time resulting from peak flow conditions may be addressed through increased dosing to achieve the required concentration and contact time to achieve the specified level of bacterial removal. Depending on the target indicator organism, UV systems may exhibit a higher sensitivity to variations in influent flow and water quality conditions relative to chlorine-based systems caused, in part, by short contact times. However, as long as a UV system design is based on a representative range of influent flow and water quality conditions, including wet weather events, UV systems can achieve disinfection performance requirements with a faster response time to changing conditions compared to chlorine-based systems.

Chlorine-based systems rely on relatively simple and low-power chemical feed equipment, whereas UV systems typically require external technical assistance for anything beyond routine maintenance and have significant power requirements that must be accommodated in a WRRF's emergency power plan. Chlorine-based systems require storage, may be handled by an operator, and present the possibility of exposure to hazardous chemicals, whereas UV systems offer a significantly lower operator hazard risk during operation and maintenance. Ultraviolet systems do not produce DBPs that are currently being restricted in some WWRF discharge permits. Overall reliability of each system is a function of site-specific conditions and operations and maintenance preferences. Advantages and disadvantages as well as operational considerations for both types of systems are summarized in Table 15.2 (Chandler et al., 2004) and detailed in other sources (Black & Veatch, 2010; WEF et al., 2010).

8.0 OTHER DISINFECTANTS

8.1 Ozone

Ozone is one of the strongest oxidants available for disinfection of wastewater and offers a number of advantages relative to other disinfectants; however, it has historically had limited application for the disinfection of wastewater (Black and Veatch, 2010; WEF et al., 2010; WERF, 2008). Because of cost and reliability issues, the use of ozone has been declining over the past 30 years; only seven out of 4450 significant WRRFs surveyed in a 2008 study used ozone (WERF, 2008). Ozone must be generated continuously on-site. When coupled with an existing use of liquid oxygen (such as a high-purity oxygen activated sludge system), ozone may have cost and operational advantages; however, the technical complexity and operations

TABLE 15.2 Wet weather UV disinfection and chlorination—advantages and disadvantages.

	UV disinfection	Chlorination
Contact time	Seconds	15 minutes at peak flow or maximum rate of pumping (if intermediate pumping is used)
Temperature and pH sensitivity	Low	High
Dosage control	Flexible	Difficult for intermittent and highly variable flow and water quality
Footprint	Smaller	Larger—storage required
Dechlorination Required	No	Yes
DBP formation	No	Yes
Chemical residual	No	Yes
Transport accessibility required	No	Yes
Cost	Construction and life cycle cost evaluation recommended (if dechlorination is required, life cycle costs for UV may be competitive pending site-specific factors)	

and maintenance requirements for operation of ozone generation systems often result in noncompetitive costs. Specific use of ozone for wet weather disinfection requires a wide operating range for the ozone generation system that may further affect cost and operational factors. Similar to UV systems, ozone systems are dependent on emergency power supply availability. Additionally, the use of ozone for disinfection requires containment of the contact chamber offgases, which may pose direct hazards to operations staff.

8.2 Chlorine Dioxide

Chlorine dioxide is created by the chemical reactions of sodium chlorite with gaseous chlorine, aqueous hypochlorite, or hydrochloric acid. Chlorine dioxide is a stronger oxidant than chlorine, but not as strong as ozone. Like ozone, chlorine dioxide requires on-site generation because of its extreme volatility. Advantages of chlorine dioxide include the following: it is unreactive with ammonia, effective over a wide range of wastewater pH, and a strong bacterial and viral oxidant. However, as of 2010, no known WRRFs or CSO facilities in the United States were using chlorine dioxide for disinfection of wastewater effluent (Black & Veatch, 2010).

8.3 Hybrid Systems

The use of a hybrid disinfection system (i.e., the use of one system for base dry weather flows and another system for treatment of wet weather flows) is a possible wet weather management strategy for WRRFs, depending on site-specific conditions, costs, and operational preferences. A case study in Section 9.1 highlights a hybrid UV/chlorination system.

8.4 Emerging Disinfectants (Specific to Wet Weather Applications)

8.4.1 Peracetic Acid

Peracetic acid (PAA) is a stronger oxidant than chlorine and chlorine dioxide, but not as strong as ozone. Peracetic acid is typically delivered to the site in totes or in bulk at concentrations of 12 to 15%. Dosages of 1 to 2 mg/L PAA are typical, but could be as high as 5 mg/L for disinfecting secondary effluent, depending on the target organisms and water quality (U.S. EPA, 2013). According to Rossi et al. (2007), a 4-log reduction of *Escherichia coli* was achieved with doses ranging from 5 to 10 mg/L and 35 to 50 minutes of contact time. Rossi et al. (2007) also indicated that disinfection efficiency improves with low TSS concentration; however, PAA can be an effective disinfectant in primary effluents and with TSS concentrations up to 100 mg/L. Significant disadvantages of PAA disinfection include potential microbial regrowth because of the presence of acetic acid, increased organics in the effluent, and a reduced disinfection efficiency against some viruses and parasites (e.g., *Cryptosporidium* and *Giardia*) (Rossi et al., 2007). With respect to wet weather applications, PAA has similar advantages and disadvantages to liquid-feed and storage chlorine-based systems.

8.4.2 Bromine

Bromo-chloro-dimethylhydantoin (BCDMH) is a crystalline substance that disassociates slowly in water to form hypochlorous acid and hypobromous acid. It is highly stable in powder form and has a storage life greater than 1 year, which makes this disinfectant attractive for periodic wet weather conditions. Bromo-chloro-dimethylhydantoin can achieve bacterial reductions comparable to sodium hypochlorite/chlorine and shorter contact times are required. Bromo-chloro-dimethylhydantoin produces DBPs and a residual that may require a quenching agent (U.S. EPA, 2013).

Similar to chlorine disinfection, bromine disinfection is affected by wastewater pH. Hypobromous acid is more predominant than hypochlorous acid at the pH of typical wet weather flows, which is typically in the neutral range; according to Boner et al. (2002), wet weather wastewater pH ranged from 6.3 to 7.4 at the Columbus, Georgia, Uptown Park CSO facility. In the presence of ammonia, bromine disinfection produces bromamines (mono- and dibromamine) much like chlorine disinfection produces chloramines; however, bromamines decay rapidly (within less than an hour), whereas chloramines require many hours to decay. The faster decay rate of bromamines results in a lower potential effect on aquatic life and human health (Boner et al., 2002).

9.0 CASE STUDIES

9.1 Hybrid System: Ozone-Enhanced Ultraviolet with Chlorination/Dechlorination (Indianapolis Belmont) (Chiu et al., 2011)

Water resource recovery facility .... Indianapolis Belmont Advanced Wastewater Treatment Plant
Location:..Indianapolis, Indiana
Disinfection process(es): Ozone-enhanced UV and chlorination/dechlorination
Average daily flow:..570 ML/d (150 mgd)
Peak flow capacity: ....1140 ML/d (300 mgd)

9.1.1 Background

The Indianapolis Belmont Advanced Wastewater Treatment Plant (Belmont facility) previously had a peak primary treatment capacity of 1140 ML/d (300 mgd); however, the peak secondary treatment and disinfection capacity were limited to 570 ML/d (150 mgd). The previous method of disinfection was chlorination/dechlorination, with sodium hypochlorite applied after secondary clarification. Contact time was achieved within the filter cells and four existing contact tanks. Sodium bisulfite was applied to the contact tank effluent before the outfall. The city of Indianapolis performed an extensive conceptual evaluation to increase the wet weather disinfection capacity to 1140 ML/d (300 mgd).

9.1.2 Alternatives

The alternatives analysis was performed to consider two disinfection processes: Technology 1 would use existing infrastructure to maintain dry weather flow capacity and Technology 2 would treat wet weather flows with new infrastructure. The schematic shown in Figure 15.1 represents

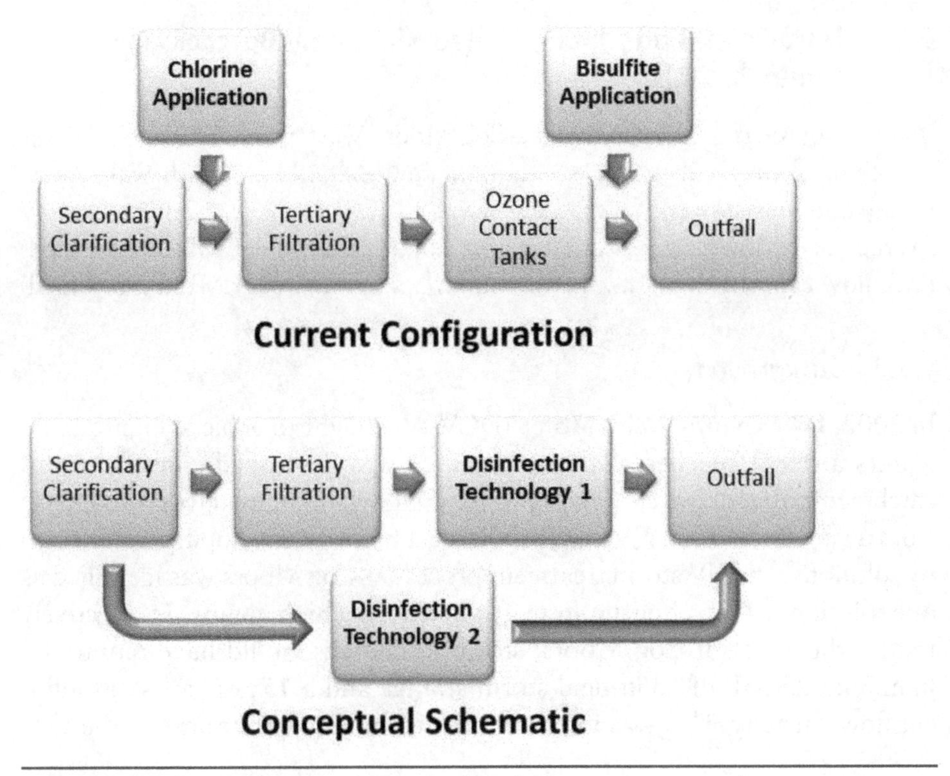

FIGURE 15.1 Schematic of current disinfection vs conceptual.

the current disinfection process compared to the conceptual evaluation of Technology 1 and Technology 2.

The concept considered for the Belmont facility expansion consisted of permutations of the following disinfection technology candidates: chlorination/dechlorination, ozonation, UV, and UV preceded by low dose ozonation.

9.1.3 Conclusions

The selected disinfection alternatives consisted of UV preceded by low-dose ozonation for Technology 1 and chlorination/dechlorination for Technology 2. Using ozonation before UV for Technology 1 improves the efficacy of disinfection by UV irradiation, enhances process reliability, allows for the reuse of existing infrastructure, reduces levels of certain microconstituents, raises effluent dissolved oxygen concentration, and does not introduce any objectionable chemicals to the treated effluent.

The selection of chlorination/dechlorination for Technology 2 evolves the existing application from a continuous process to an intermittent process. This decreases the sodium hypochlorite and sodium bisulfite usage, the associated operational hazards, and the amount of chlorination byproducts present in the treated effluent.

9.2 Ultraviolet (Cog Moors, United Kingdom) (Barcock and Scannell, 2010)

Water resource recovery facility: ..... Cog Moors Wastewater Treatment Works
Location: .... South Wales, U.K.
Disinfection process(es): .... UV
Average daily flow: .... 86.4 ML/d (22.8 mgd)
Peak flow capacity: .... 362.4 ML/d (95.8 mgd)

9.2.1 Background

In 2002, Dwr Cymru Welsh Water (DCWW) sought to achieve high-quality waters at local beaches. A design strategy was developed considering all catchments affecting the receiving waters. Models were used to assess effects from a variety of DCWW assets, providing a basis for developing water quality solutions. The UV storm treatment process at Cog Moors was identified as one solution within a holistic strategy to achieving high quality. To effectively reduce the effect of Cog Moors, storm discharges would have required a minimum 25 ML of additional storm storage and a 15% increase in influent flow capacity. This was impracticable within the constraints of the site.

9.2.2 Analysis

A UV storm treatment process, capable of delivering a 2-log reduction in bacteria, was deemed the most effective solution and provided other benefits such as a small footprint, low carbon footprint (construction and operation), low construction and operational cost, treatment of all storm flows, bacteria reduction, and operational flexibility.

The UV system that was selected to achieve project objectives was a Trojan UV4000PLUS packaged UV stormwater treatment system. The facility comprises three parallel 103-ML/d (27-mgd) capacity channels, each enclosing five modules of 22 medium-pressure lamps and an array of static mixers that prevent the formation of blind spots. The full installation provides a duty/duty/standby arrangement. The rate of flow through the duty channels is equalized and monitored by level sensors. The UV dose is controlled by flowrate, transmissivity, and intensity recorded in real time by in situ sensors. The minimum applied power rating is 30% of full rating. In the event of a control failure, the UV lamps are operated at full power rating.

9.2.3 Conclusions

During wet weather events, the existing storm tanks operated as normal. As the storm tanks overflow, the UV lamps are activated and the storm tank is diverted to the UV facility. Using the existing storm tanks provides initial settlement of the storm effluent and increasing transmissivity, thereby requiring less power consumption by the UV lamps.

Since commissioning, the UV facility has operated successfully. Routine monitoring of effluent quality has shown that the facility meets, and frequently exceeds, the required log reduction in bacteria.

9.3 Blended Effluent Bulk Hypochlorite Chlorination (Boczek et al., 2008)

Water resource recovery facility: .... Various (3)
Location: .... New York, New York
Disinfection process(es): .... Chlorination/Dechlorination
Average daily flow: .... 450 ML/d (120 mgd)
.... 230 ML/d (60 mgd)
.... 1040 ML/d (275 mgd)
Peak flow capacity: .... 900 ML/d (240 mgd)
.... 460 ML/d (120 mgd)
.... 2080 ML/d (550 mgd)

9.3.1 Background

Blending is the practice of diverting a part of peak wet weather flows at WRRFs, after primary treatment, around biological treatment units and combining effluent from all processes before disinfection and subsequent discharge from a permitted outfall. The practice of blending has been the subject of administrative and legal review.

This study was performed by U.S. EPA on three WRRFs in New York. The project's intent was to determine the microbiological effect of blending primary effluent flows that are in excess of secondary treatment capacity with the secondary effluent before disinfection at large municipal WRRFs (U.S. EPA, 2010).

9.3.2 Analysis

Samples of primary and undisinfected secondary effluents were collected from three local conventional activated sludge WRRFs. These effluents were then mixed thoroughly, but gently at three different ratios (1:9, 3:7, and 5:5) to produce blended effluents. Samples of each of these blended products were assayed for indicator organisms, suspended solids, turbidity, total organic carbon, ammonia, redox potential, pH, and temperature. Blended effluents were then mixed with a dilute sodium hypochlorite solution to a final total chlorine concentration of approximately 2.0 and 5.0 mg/L and sampled after 30 minutes of contact time. Sodium thiosulfate was then added to the sterile sample vessel to immediately quench chlorine oxidation. Finally, samples were assayed for microorganisms.

9.3.3 Conclusions

The data showed that chlorination of blended effluents was highly effective in reducing the numbers of the most common indicator, fecal coliforms. It also demonstrated that the increased amount of primary effluent reduces the disinfectant's ability to kill or inactivate the coliphage population, suggesting that other viruses may also survive chlorination. It is likely that disinfection of the blended wastewater effluents was affected not only by the chemical properties, but by the physical properties as well. Particulate matter from primary effluents provides a potential way to limit direct contact of disinfecting agents with microorganisms. Such a shielding effect may be more significant for virus particles than for bacteria.

9.4 Alternate Disinfectant: Peracetic Acid (LaGorga and Cameron, 2011)

Water resource recovery facility: ....Combined Sewer Overflow Outfall No. 002
Location: .... Oswego, New York
Disinfection process(es): .... PAA
Average daily flow: .... N/A
Peak flow capacity: ....235 ML/d (62 mgd)

9.4.1 Background

The city of Oswego's westside wastewater collection system covers an area of approximately 405 ha (1000 ac), of which approximately 140 ha (350 ac) are composed of combined sewers. Combined Sewer Overflow Outfall No. 002 is the only permitted CSO outfall on the city's west side.

The city's goal was to eliminate or significantly reduce the size of the disinfection contact tank, thereby reducing the life cycle cost of a combined wastewater disinfection facility.

9.4.2 Analysis

A bench-scale disinfection pilot study was completed to test the effectiveness of PAA and sodium hypochlorite at high doses and short contact times. The effluent goal for the disinfection pilot study was a fecal coliform density of 200 counts/100 mL.

9.4.3 Conclusions

The results indicated that the biocide effect of PAA against fecal coliforms was similar to that of sodium hypochlorite. The data showed that significant reductions in fecal coliform (4-log) can be achieved at short contact times (1 and 2 minutes) with high doses (30 mg/L) of disinfectant.

9.5 Alternate Disinfectant: Bromine (Boner et al., 2002)

Water resource recovery facility: .... Uptown Park CSO Treatment Facility
Location: .... Columbus, Georgia
Disinfection process(es): .... Bromine (BCDMH)
Average daily flow: .... N/A
Peak flow capacity: .... 182 ML/d (48 mgd)

9.5.1 Background

The Columbus Water Works in Columbus, Georgia, included programs for demonstrating wet weather technologies in their $95 million CSO control program. The programs evaluated various disinfection techniques, including bromine as BCDMH (1-bromo, 3-chloro, 5,5-dimethylhydantoin).

9.5.2 Analysis

Multiple flow-through technologies were tested side-by-side to evaluate performance, operation, and costs for application to control CSOs. Technology applications included chemical disinfection alternatives, specifically bromine as BCDMH, a relatively new disinfectant.

9.5.3 Conclusions

Several advantages exist for BCDMH over other chemicals for CSO disinfection. This chemical disassociates into greater proportions of the active hypohalous acid at CSO pH levels. When combined with ammonia, BCDMH appears to be 3 times more effective as an oxidant and germicide than the other disinfectants tested and has the advantage of being less stable and quick to decompose, leaving little or no residual.

Bromine's ability to be more effective under the quality variations present in CSOs and stormwater runoff and its decomposable nature to rapidly dissipate may identify it as a viable solution to wet weather disinfection.

10.0 REFERENCES

Barcock, N.; Scannell, C. (2010) Cog Moors Storm UV Disinfection: Sustainability through New Technology and Seasonal Consenting. *Wastewater Treat. Sewerage*, 2010, 51–54.

Bell, K. Y.; Parke, S.; Dillon, J.; Sun, J. W. (2011) Wastewater Process Modifications for Addressing TSS to Improve UV Disinfection. *Proceedings of the 84th Annual Water Environment Federation Technical Exhibition and Conference* [CD–ROM]; Los Angeles, California, Oct 15–19; Water Environment Federation: Alexandria, Virginia; pp 110–120(3).

Birkel, P. F.; Firmin, S.; Fluet, R. (2007) Hypochlorite Conversion in Portland, ME Compliance, Wet Weather Disinfection, Improved Automation, and Success. *Proceedings of the 80th Annual Water Environment Federation Technical Exhibition and Conference* [CD–ROM]; San Diego, California, Oct 13–17; Water Environment Federation: Alexandria, Virginia; pp 175–192(1).

Black & Veatch (2010) *White's Handbook of Chlorination and Alternative Disinfectants*; Wiley & Sons: Hoboken, New Jersey.

Boczek, L. A.; Johnson, C. H.; Meckes, M. C. (2008) Chlorine Disinfection of Wet Weather Managed Flows. *Proceedings of the 81st Annual Water Environment Federation Technical Exhibition and Conference* [CD–ROM]; Chicago, Illinois, Oct 18–22; Water Environment Federation: Alexandria, Virginia; pp 7675–7678(7).

Boner, M.; Kim, J. Y.; Muller, R. J. (2002) Is Bromine Disinfection a Viable Wet Weather Solution. *Proceedings of the 75th Annual Water Environment Federation Technical Exhibition and Conference* [CD–ROM]; Chicago, Illinois, Oct 28–Sept 2; Water Environment Federation: Alexandria, Virginia; pp 275–296(1).

Chandler, K. L.; Steidel, R.; Watson, C.; Maisch, F. E.; Cronin, E. J.; Liang, L, Guhse, G. L. (2004) Richmond, Virginia Investigates UV Disinfection of Large CSO Outfall. *Proceedings of the 77th Annual Water Environment Federation Technical Exhibition and Conference* [CD–ROM]; New Orleans, Louisiana, Oct 2–6; Water Environment Federation: Alexandria, Virginia; pp 420–443(15).

Chiu, H. H.; Reichlin, D. B.; Erhardt, R. S.; Jousset, S. (2011) Ozone-Enhanced UV Disinfection at the Indianapolis Belmont Advanced Wastewater Treatment Plant. *Proceedings of the 84th Annual Water Environment Federation Technical Exhibition and Conference* [CD–ROM]; Los Angeles, California, Oct 15–19; Water Environment Federation: Alexandria, Virginia; pp 385–393(3).

Fitzpatrick, J.; Hunter, G.; Phillips, H.; Williamson, B. (2007) Wet Weather Disinfection Along the Muddy Missouri . . . Piloting for Success. *Proceedings of the 80th Annual Water Environment Federation Technical Exhibition and Conference* [CD–ROM]; San Diego, California, Oct 13–17; Water Environment Federation: Alexandria, Virginia; pp 193–222(1).

Great Lakes–Upper Mississippi River Board of State and Provincial Public Health and Environmental Managers (2004) *Recommended Standards for Wastewater Facilities;* Health Research, Inc.: Albany, New York.

Hunter, G. (2012) Impact of Upstream Process on Meeting Disinfection Limits with UV Light. *Proceedings of the 85th Annual Water Environment Federation Technical Exhibition and Conference* [CD–ROM]; New Orleans, Louisiana, Sept 29–Oct 3; Water Environment Federation: Alexandria, Virginia; pp 7961–7970.

LaGorga, J.; Cameron, R. (2011) Pushing the Limits of Wet-Weather Flow Disinfection. *Proceedings of the 80th Annual Water Environment Federation Technical Exhibition and Conference* [CD–ROM]; San Diego, California, Oct 13–17; Water Environment Federation: Alexandria, Virginia; pp 823–830(5).

Metcalf & Eddy, Inc. (2003) *Wastewater Engineering: Treatment and Reuse,* 4th ed.; McGraw-Hill: New York.

Muller, J.; Lem, W. (2011) UV Disinfection of Stormwater Overflows and Low UVT Wastewaters. *IUVA News,* **13** (3), 13–17.

Muller, J.; Petri, B. (2013) Trojan Technologies. Wet Weather Considerations for UV Disinfection Systems. Personal communication.

National Water Research Institute (2003) *Ultraviolet Disinfection: Guidelines for Drinking Water and Water Reuse;* National Water Research Institute: Fountain Valley, California.

Rossi, S.; Antonelli, M.; Mezzanotte, V.; Nurizzo, C. (2007) Peracetic Acid Disinfection: A Feasible Alternative to Wastewater Chlorination. *Water Environ. Res.,* **79** (4), 341–350.

Scheible, O.; McGrath, J. (2000) Generic Verification Protocol for High-Rate, Wet-Weather Flow Disinfection Applications. U.S. EPA Environmental Technology Verification Program: Washington, D.C.

Ting, L. (2012) The Effect of Upstream Processes on UV Disinfection and Chlorination in the Greater Cincinnati Sewer District. *Proceedings of the 85th Annual Water Environment Federation Technical Exhibition and Conference* [CD–ROM]; New Orleans, Louisiana, Sept 29–Oct 3; Water Environment Federation: Alexandria, Virginia; pp 6187–6192.

U.S. Environmental Protection Agency (1999a) *Combined Sewer Overflow Technology Fact Sheet: Alternative Disinfection Methods;* EPA-832/F/99-033; U.S. Environmental Protection Agency, Office of Water: Washington, D.C.

U.S. Environmental Protection Agency (1999b) *Wastewater Technology Fact Sheet: Chlorine Disinfection,* EPA-832/F-99-06; U.S. Environmental Protection Agency, Office of Water: Washington, D.C.

U.S. Environmental Protection Agency (2002) *Environmental Technology Verification Report: Performance of Induction Mixers for Disinfection of Wet Weather Flows;* NSF 02/02/EPA WW399; U.S. Environmental Protection Agency: Washington, D.C.

U.S. Environmental Protection Agency (2005) Proposed EPA Policy on Permit Requirements for Peak Wet Weather Discharges from Wastewater Treatment Plants Serving Sanitary Sewer Collection Systems. http://cfpub.epa.gov/npdes/wetweather.cfm (accessed May 2014).

U.S. Environmental Protection Agency (2007) *Wastewater Management Fact Sheet: In-Plant Wet Weather Peak Flow Management;* EPA-832/F-07-016; U.S. Environmental Protection Agency, Office of Water: Washington, D.C.

U.S. Environmental Protection Agency (2010) *Impact of Wet-Weather Peak Flow Blending on Disinfection and Treatment: A Case Study at Three Wastewater Treatment Plants;* EPA-600/R-10/003; U.S. Environmental Protection Agency, Office of Research and Development: Washington, D.C.

U.S. Environmental Protection Agency (2013) *Emerging Technologies for Wastewater Treatment and In-Plant Wet Weather Management;* EPA-832/R-12-011; U.S. Environmental Protection Agency, Office of Wastewater Management: Washington, D.C.

Water Environment Federation (2013) *Guide for Municipal Wet Weather Strategies;* Water Environment Federation: Alexandria, Virginia.

Water Environment Federation; American Society of Civil Engineers; Environmental & Water Resources Institute (2010) *Design of Municipal Wastewater Treatment Plants:* 5th ed.; WEF Manual of Practice No. 8; ASCE Manuals and Reports on Engineering Practice No. 76; Water Environment Federation: Alexandria, Virginia.

Water Environment Research Foundation (2002) *Best Practices for Treatment of Wet Weather Wastewater Flows;* WERF 00-CTS-6; Water Environment Research Foundation: Alexandria, Virginia.

Water Environment Research Foundation (2008) *Disinfection of Wastewater Effluent—Comparison of Alternative Technologies;* WERF 04-HHE-4; Water Environment Research Foundation: Alexandria, Virginia.

Water Environment Research Foundation (2009) *Characterizing the Quality of Effluent and Other Contributory Sources during Peak Wet Weather Events,* Final Report; Project 03-CTS-12PP/PPa; Water Environment Research Foundation: Alexandria, Virginia.

Wood, P.; Hunter, G.; Kobylinski, E. (2005) To PLC or not to PLC UV Systems—Alternatives for UV System Control. *Proceedings of the Water Environment Federation: Disinfection*; Water Environment Federation: Alexandria, Virginia; pp 582–591.

16

Residuals

Miguel Vera; Stephanie Spalding; and Scott Phipps, P.E.

1.0 EFFECTS OF WET WEATHER EVENTS IN SLUDGE HANDLING

1.1 Introduction

Sludge composition and concentration are highly dependent on the unit operations that are part of the pretreatment and primary and secondary treatment processes of water resource recovery facilities (WRRFs) and on the effectiveness of the solids/liquid separation equipment installed. Screening and grit removal are of special interest during wet weather events because those two processes are susceptible to being overloaded in the initial flow surge or first flush. Chapter 11 discusses in detail the wet weather related considerations of managing grit and screenings at a WRRF. This chapter focuses on related considerations for the residual sludge streams also generated in WRRFs.

1.1.1 First Flush Effects

First flush affects the sludge composition as larger quantities of grit are allowed to enter into the WRRFs. As much as 70 to 90% of the annual grit load can be received at the WRRFs during a handful of first flush events (Osei and Andoh, 2008; Wilson, 1998). The initial flush grit load results in an increased amount of inert material in the sludge and lower volatile suspended solids (VSS)/total suspended solids (TSS) ratios, which will strain settling and thickening equipment. Wet weather events pass between 60 and 80% of the total mass of TSS conveyed to the facilities within approximately the first 30% of the flow (Figure 16.1).

1.1.2 Sludge Variations in Diverse Flow Sheets

Primary sludge is reported to have a wide range of VSS concentrations, from 64 to 93% with a typical value of 77%, in which a good degritting system

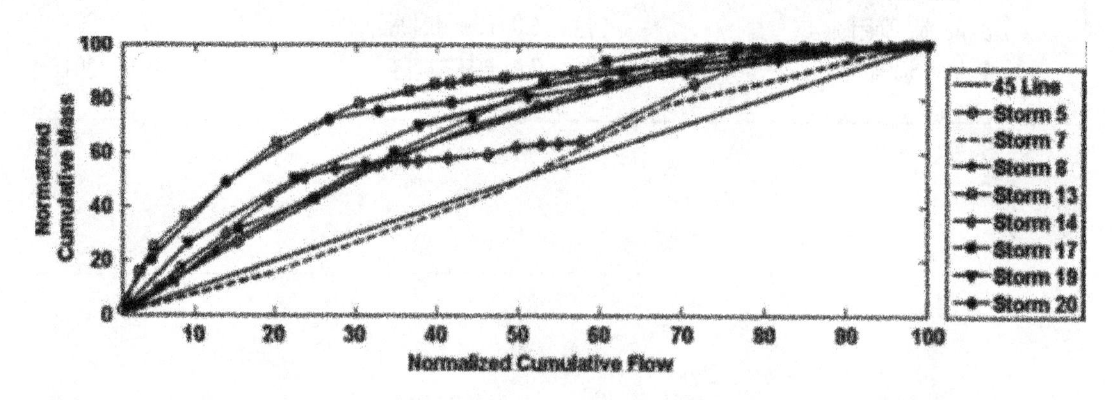

FIGURE 16.1 Cumulative mass first flush vs cumulative flow (Barco et al., 2008).

is in place. In contrast, the VSS value for primary sludge during a severe storm is only reported at 40% (U.S. EPA, 1979; WEF et al., 2010).

Facilities with inadequate grit removal can also have increased primary sludge production in which lower VSS concentrations (60% or less) reflect a higher proportion of inorganics and grit in the primary solids. Sludge degritting is an option that can help facilities deal with this use; further discussion on its application is found in Chapter 11.

Similarly affected are the sludge yields in the suspended growth processes. Net sludge production yield factors are reported in facilities using primary clarifiers varying from 0.33 to 0.88 mg TSS/mg 5-day biochemical oxygen demand (BOD_5), whereas, at sites without primary sedimentation, the range is 0.62 to 1.18 mg TSS/mg (BOD_5 for sludge ages up to 30 days at three temperatures: 10, 20, and 30 °C [50, 68, and 86 °F] Droste, 1997). These values will be reduced during wet weather events as the inorganic content will increase initially after the first flush and then the dilution effect will be present to lower the sludge concentration. Sludge production and yields are discussed elsewhere in greater detail for both primary and secondary sludge (Dold, 2007).

As discussed in Chapter 12, chemically enhanced primary treatment (CEPT) can be used to handle wet weather events as it can improve the removal efficiencies for suspended solids, settleable solids, biochemical oxygen demand (BOD), and chemical oxygen demand (COD) while being able to operate at higher surface overflow rates (Kruger et al, 2005). It is important to consider that the addition of CEPT to existing facilities will result in greater quantities of solids that must be managed (approximately 15 to 20%), and the solids tend to have a lower concentration (2.0% for iron dosed sludge) than those from conventional primary sludge (2 to 7%) (U.S. EPA, 1979).

1.2 Thickening

Solids concentrations in the sludge feed to thickening processes will be reduced while also having a greater component of grit, which will cause higher and even excessive wear in the equipment, either in the gravity belt thickener belts or in the centrifuge bowls.

Facilities using CEPT or chemical precipitation of phosphorus have effectively used flocculants such as aluminum sulfate [$Al_2(SO_4)_3$], ferric sulfate [$Fe_2(SO_4)_3$], PAC, and PAFC. These poly-metal salt flocculants (PAC and PAFC) had been reported to have a more effective removal efficiency of pollutants than metal salt ones [$Al_2[SO_4]_3$ and $Fe_2[SO_4]_3$], which means that the former dosage is less than the latter with the same removal efficiency of pollutants. Additionally, flocculant aids such as PAM, PVA, and AS also have been used to enhance the coagulation–flocculation process at relatively low coagulant dosage. The use of any of these chemicals will increase the amount of primary solids and will also tax the capacity of exiting equipment if not designed for the additional loads and/or flows. Furthermore, the sludge produced will vary with not only the addition of the different chemicals for coagulation, but also with the addition of polymers for flocculation. The dual conditioning use of coagulants and polymer will favor better thickening than a single conditioner, either coagulant or polymer, alone. However, this only holds to a certain point as higher concentrations of coagulant lead to thickening inhibition or blinding of belts in the thickening equipment (Kozak et al., 2011) and, for the most part, with sludges that are fresh, which is the case for most wet weather operations. As in all conditioning operations, optimal dosage determination is a local practice that needs to be done with jar testing.

1.3 Dewatering

In dewatering operations in wet weather events, the solids will have shorter sludge ages and often will be in a diluted state. This will minimize the effectiveness of dewatering because the larger flowrates will force the operators to process as much sludge flow as possible during the event. The dewatering equipment (belt press, centrifuge, screw press, etc.) capacity needs to be designed if a new facility is being considered or if an existing facility is to be evaluated in light of the levels of processing required for the wet weather events expected.

In the case of a belt filter press type, this dewatering equipment should have adequately sized gravity zone filtration areas to handle these dilute flows. Not all equipment is equipped in this manner, but there is a belt filter press dewatering machine in the market that is especially suited for optimization of the gravity zone filtration and handling dilute flows because it features a three-belt design with an independent gravity zone that is decoupled from the dewatering zone, as depicted in Figure 16.1.

The separation of the gravity zone and dewatering zone in this belt press allows the operator to maximize the removal of water in a thin sludge while still pressing at the best conditions to maintain a high cake dryness output as much as possible.

1.4 Pumping

Pumping of sludge in a wet weather event presents the challenge of having a high massic load at the initial flows, and, later, a dilute high-flow condition. In either case, primary sludge pumps, return activated sludge pumps, and sludge thickener pumps need to be sized properly and with a piping system that can accommodate the range of flows expected for the dry weather conditions while maintaining minimum velocities to prevent settling in the pipe and avoid septicity. At the same time, the pumps should handle the wet weather flows within maximum velocities that do not cause headloss issues or piping deterioration.

1.5 Storage

Intermediate storage such as thickened, blended, or holding tanks before stabilization or digested storage tanks and redundant or secondary digesters after stabilization are wide posts on the processing that allow operators to provide the additional storage volume necessary to handle wet weather events. These tanks should be set at their lower practical liquid levels in anticipation of the events.

Storage facilities for sludge that is disposed of regularly and not stored between stabilization and land application cycles are typically sized for at least 5 to 7 days of sludge volume to handle long weekends, holidays, or other maintenance disruptions. If thickening and dewatering facilities have been sized for a single shift operation, then wet weather operations benefit from running the equipment on a continuous basis until the wet weather event flows subside. In facilities where redundant equipment is available, putting all units into service will help to handle the wet event flows. Depending on the cost of equipment, energy, and chemicals, the decision to provide increased storage to handle wet weather sludge flows (or to provide sludge processing equipment redundancy or both) needs to be analyzed in each case. In many operations, preserving byproducts of the solids handling process (e.g., biogas, compost, or pelletized biosolids that may provide a source of income to the facility) are critical.

For facilities that store the biosolids in between stabilization and land application cycles, the storage volume needed would depend on agricultural needs and climatic issues, with the storage requirement typically being much larger that the volume discussed previously.

2.0 EFFECTS OF WET WEATHER EVENTS IN STABILIZATION PROCESSES

2.1 Preparation of Process Facilities before a Wet Weather Event

The development of a wet weather operating plan for WRRFs that include combined sewers is mandated by regulations enacted in 1994 in the federal combined sewer overflow control policy. However, it is not mandated for sites that only receive flow from sanitary sewers because the policy is still in draft form and, as a result, a few states implement a requirement for it, but it is not a widespread practice. It is recommended to develop a wet weather operation plan regardless of the source of the flows being treated. The process of preparing this document will provide operations personnel with the tools to identify areas of improvement and potential solutions to those bottlenecks in their processes and equipment. The policy and process to prepare this document is described in U.S. Environmental Protection Agency's Combined Sewer Overflow Control Policy published in 1994 (U.S. EPA, 1994).

A good strategy to prepare the solids handling facilities before wet weather season or events is to ensure that the following activities are completed:

- Verify that maximum solids storage capacity is available,
- Ensure all solids processing equipment is in operating condition,
- Investigate/prepare alternative solids disposal routes for wet weather events, and
- Reduce solids volume in digesters before wet weather events.

Additionally, provisions need to be made to ensure that emergency power generation equipment is in good operational condition before a wet weather event because electric grid power accessibility might be reduced or completely cut off because of weather-related incidents.

2.2 Anaerobic Digestion

Anaerobic digestion is a process in which the performance is highly dependent on having its operating parameters within the appropriate ranges of temperature. Wet weather events can have a significant effect on the digester operating parameters because larger flows of sludge to the biosolids handling facilities will lower the temperature in the digester, create lower organic loading, and require more heating energy to maintain either the mesophilic (35 to 39 °C [95 to 102 °F]) or thermophilic (typically between 50 and 57 °C [122 and 135 °F]) temperature range. Temperature changes greater than 1 °C/d (1.8 °F/d) can result in process failure (WEF et al., 2010).

The best defense for sudden temperature drop resulting from receiving dilute sludge in a digester is to have thickening and/or storage facilities with the capacity to handle the extra flows generated by the wet weather flows.

The solids retention time (SRT) (or hydraulic retention time, if no recycling or decanting is done) is directly related to the effectiveness of each of the anaerobic digestion process steps and, as such, increasing flows in a digester means a reduction of treatment time and, possibly, loss of biomass. The most common effect of shorter SRTs is the drop in gas production, which, when combined with a sudden decrease in temperature, may affect the classification of the biosolids being produced, especially for facilities that operate under a time and temperature scheme to meet Class A biosolids performance. If the goal is to meet the Class A biosolids performance, a storage strategy for sludge flows above the level of processing of their thickening facilities is recommended.

Co-generation facilities operations will be altered and possibly shut down by the reduction in biogas production because the pressure in the biogas storage unit will continue to decrease if the co-generation system is in operation without having the biogas replenish quickly enough. Additionally, there will be a lower amount of available energy for immediate use at the WRRF.

2.3 Aerobic Digestion

Aerobic digestion is a long-term aeration process that is used to stabilize and reduce the total mass of organic waste by biologically destroying volatile solids. The three main processes used in the United States currently and in the recent past are (1) conventional aerobic digestion (CAD), (2) autothermal thermophilic aerobic digestion (ATAD), and (3) an improvement of CAD designated as *prethickened aerobic digestion* (PAD).

Typically, aerobic digestion in any of these configurations is used mostly to stabilize only waste activated sludge in facilities that do not produce primary sludge; however, the implementation of more stringent nutrient removal permits throughout the country has increased the interest and use of aerobic digestion to stabilize mixtures of both primary and secondary sludge because this process will not have a side stream return with as much phosphorus or ammonia as that of an anaerobic digestion process.

Conventional aerobic digestion is highly susceptible to wet weather events because it uses sludge that is typically not thickened beyond the settling on the secondary clarifier or minimally thickened by decanting up to 1.5 to 2.0%. Both ATAD and PAD processes are also highly dependent on the thickening, with typical feed concentrations greater than 4% and 3 to 4%, respectively (WEF et al., 2010), to be able to generate adequate retention times that can maintain the process within the recommended operating guidelines.

Using storage and holding tanks before thickening is a desirable strategy for facilities encountering wet weather events to avoid (1) diluting the digester contents and (2) lowering its temperature.

Another significant consideration for both CAD and PAD processes are the changes in operating temperatures related to ambient temperatures because most of these aerobic digestion systems are open tanks. In a cold climate, however, it is necessary to implement using covers to be able to operate within the design relationship of SRT and temperature, which is critical to achieving the desired level of volatile solids reduction for the stabilization of the biosolids handled in these processes.

2.4 Lime Stabilization

Traditional lime stabilization (liquid lime) can be used to meet Class B pathogen reduction requirements, whereas advanced alkaline stabilization technologies (typically using dry lime) are used to meet either Class B or Class A pathogen reduction requirements. Minimum lime doses of 25 to 40% (on a dry-weight basis as calcium hydroxide) typically are required for liquid lime Class B stabilization, and minimum doses of 15 to 30% (on a dry-weight basis as calcium hydroxide) typically are required for effective dry lime stabilization.

The lime facilities that are using the traditional liquid lime process do not have much susceptibility to the lower sludge concentrations that are present during a wet weather event because the liquid lime requirements are more closely related to the feed cake's total mass than to its volume when its solids concentration ranges from 0.5 to 4.5% (U.S. EPA, 1979).

However, although the quicklime requirement for Class B stabilization theoretically increases as the solids concentration increases, the quicklime requirement for Class A stabilization decreases as the solids concentration increases because lime is used to heat the cake to achieve Class A disinfection (a lower solids concentration will mean that more mass of water needs to be heated using quicklime to the required temperature); conversely, lime is used to raise pH for Class B (WEF et al., 2010).

In the overall facility sizing, the quantity and characteristics of the solids to be processed determine the overall sizing of the alkaline treatment equipment. Thus, the effects of a dilute incoming feed in a wet weather event will be a probable cause for variable performance of the thickening and/or dewatering processes, which are an important consideration when sizing the alkaline stabilization process because poor dewatering performance significantly increases its equipment sizing.

The size of storage facilities should take into consideration the volumes of sludge and chemical required during wet weather operation, especially for locations in which intermediate storage is used in some of the advanced alkaline stabilization processes that require a heating step to achieve Class A stabilization requirements.

Final storage of the alkaline-stabilized product is also an important design consideration. Sufficient storage should be provided if product curing is required and to accommodate road and weather conditions and fluctuations in the product marketing and distribution schedule. Adequate storage must be provided to accommodate road and weather conditions and to allow peak production for the longest anticipated time without access to the final end use location or market.

2.5 Composting

Composting is a stabilization process that prepares raw, digested, or chemically stabilized solids for use as a soil conditioner. Composting can be performed outdoors in most climates, although, to operate the process more efficiently, control odors, and reduce operating costs, many facilities are constructed under cover or in mechanized "invessel" systems to prevent rain or snowfall from increasing moisture and decreasing temperatures in the pile. In addition, covering compost piles should protect water quality by preventing runoff.

The processes most commonly used in composting biosolids are the following:

- Windrow,
- Aerated static piles, and
- Invessel systems.

Windrow composting is a process in which the biosolids/bulking agent mixture is placed in long open-air piles and turned frequently to provide aerobic conditions and to ensure that all parts of the pile are exposed to temperatures that will kill pathogens. This relatively "low-tech" process has seasonal limitations, primarily because it does not work as well in cold or wet weather.

Aerated static systems are designed so that aerobic conditions are maintained by blowing or drawing air through the compost piles by way of perforated pipes. The piles are typically insulated with finished compost or a bulking agent to maintain temperatures and can function year round in various climates or weather conditions when designed properly.

Invessel composting systems are located in completely enclosed buildings in which environmental conditions are closely monitored and controlled, which help ensure a year-round, consistent, and marketable product.

The primary controls that govern the composting process are the following:

- Properties of the initial mix,
- System aeration, and
- Detention time and mixing.

The initial mixture of biosolids and amendment or bulking agent is extremely important because an initial mixture with an adequate total solids concentration will provide porosity for proper air distribution and structural integrity for pile construction. For a good process performance, the dewatered cake should contain between 14 and 30% solids, which are then blended with drier materials (e.g., wood chips, sawdust, shredded yard waste, and ground pellets) to achieve a solids content of approximately 38 to 45%. It is at this point that the wet weather event operations need to be carefully planned because the use of bulking agents may increase because of lower cake solids in the dewatered biosolids. Preparations for the wet weather season require stockpiling bulking and amendment agents in necessary quantities and under the right storage conditions to have an operational facility during and after a wet weather event.

The compost system aeration provides oxygen for microbiological growth, removes moisture, and controls temperature. The initial design and operational guidelines for wet weather must include enough aeration capacity to operate in lower ambient temperatures and higher moisture conditions of the wet weather season to still maintain reasonable operating pile temperatures to continue the composting process.

Minimum solids retention of 60 days is recommended to produce stable compost. Composting systems typically provide from 18 to 24 days in the active composting process, followed by 30 to 45 days of compost curing (National Biosolids Partnership, 2005).

It is in the curing phase of the process that compost increases in total solids concentration; therefore, these piles and the storage piles need to be protected during a wet weather event. Temporary cover may be needed and, if left in the open, provisions need to be made for appropriate drainage and catchment of the runoff.

2.6 Thermal Processing

Thermal drying involves applying heat to evaporate water from biosolids. Thermal drying reduces the moisture content to a level below that achievable by conventional mechanical dewatering methods, minimizing the volume of the final product, and provides further pathogen reduction to Class A biosolids.

The drying rate of biosolids depends on the internal mechanism of liquid flow and the external mechanism of evaporation. Both mechanisms occur simultaneously, and either mechanism may limit the drying rate.

The three general stages of thermal drying are as follows:

1. Warm-up stage, in which steady-state conditions are reached. Limited drying occurs in this stage.

2. Constant-rate stage, in which evaporation of the interior moisture starts and typically is the longest stage, resulting in most of the drying. In this stage, the drying rate depends on the heat transfer coefficient, the area exposed to the drying medium, and, more critically for wet weather operations, the temperature and humidity difference between the drying medium and the wet surface of the sludge. It is imperative to maintain the highest level of cake dryness in the feed to the dryer because higher water content will directly increase its consumption of energy.
3. Falling-rate stage, in which the external moisture evaporates faster than it can be replaced by internal moisture. As a result, the drying rate decreases, although most of the water should have been removed at this stage.

To control the process steps, continuous monitoring of temperature, feed rates, air rates, and drying time are recommended to reduce fire hazards and to optimize operation, which can be of great help when dealing with water-laden sludge streams being produced during a wet weather event. Most designs include instrumentation packages in which operators can easily observe and respond to process indicators or alarms.

Thermal drying systems can be grouped into direct and indirect categories. Discussions of these systems can be found in literature by Metcalf & Eddy (2003) and WEF et al. (2010). Relevant and of note for wet weather operation are the following drying process components or steps:

1. Storage—storage before and following the drying process is an important system element. Sufficient storage is required upstream of the drying system to accommodate drying-system shutdowns and higher volumes of cake resulting from wet weather operations and to attenuate variations in solids production.
2. Feeder equipment—wet cake requires uniform feeding to maintain effective drying. In wet weather operations, the consistency of the sludge may vary and a self-cleaning feeder is needed to effectively control solids buildup on the feed conveyors.
3. Mixing and agitation equipment—mixing and agitation improves the rate of drying and increases the efficiency of the unit. In wet weather operations, uniform average feed conditions are difficult to maintain and appropriately sized and heavy-duty equipment is needed to break up the clumps of sludge and reduce imbalance and bearing problems that result from fouling of the process units.

3.0 EFFECTS OF WET WEATHER EVENTS IN REUSE AND DISPOSAL OUTLETS

3.1 Land Application

Wet weather conditions may hinder timely or uninterrupted biosolids disposal or utilization. Although rules for application vary from state to state, sludge typically must not be applied to land during precipitation events or when the soil is saturated with water. It is also recommended that operators avoid applying sludge if a 6.35-mm (0.25-in.) rainfall event occurs during the 24 hours before the scheduled time of application or when precipitation is imminent and there is a risk that the land-applied waste will run off to wetlands or other receiving waters.

Land application in wet weather conditions will be impractical because trucks and equipment can get bogged down under the soggy and muddy conditions of an application site. If there are enough of these occurrences during a year, it is recommended that storage areas be provided that can receive and hold the biosolids until the application site is capable of receiving the biosolids.

3.2 Landfill

Landfills have operational challenges during and after wet weather events that severely limit their availability for disposal of sludge or biosolids. At a minimum, wet conditions will increase round trip times for disposal trucks because the disposal times in the active landfill will take longer; eventually, the landfill will stop receiving loads resulting from their receiving area being filled because the working or active face is no longer reachable or the road to the site is no longer accessible.

The frequency of wet weather events and local conditions of the landfill site will dictate the need for alternative disposal routes and/or storage options and, therefore, optional sites need to be included as part of a wet weather operations plan.

4.0 EFFECTS OF WET WEATHER EVENTS IN SIDE STREAM TREATMENT

4.1 Side Stream Treatment Options

There are several alternatives for side stream treatment, as delineated in Table 16.1

TABLE 16.1 Side stream options (adapted from Law and Stinson, 2012).

Sidestream treatment options	
Biological—nitrogen removal	**Physicochemical—nitrogen and phosphorus removal**
1. Nitrification/denitrification and bioaugmentation With RAS and SRT control With RAS Without RAS	1. Ammonia stripping Steam Hot air Vacuum distillation
2. Nitritation/denitritation Chemostat SBR Postaerobic digestion	2. Ammonia recovery Ion exchange
3. Deammonification Suspended growth sequencing batch reactor Attached growth moving bed biofilm reactor/integrated fixed film activated sludge Upflow granular process	3. Phosphate recovery Fluidized bed reactor Aerated reactor

4.2 Preparation of Facilities before a Wet Weather Event

Several of the processes presented in Table 16.1 present operation schemes that feature small hydraulic retention times, high temperature, and batch operation (Constantine and Katehis, 2007). All of these characteristics create operational challenges when faced with the higher flows of a wet weather event. Each process needs to be evaluated in light of not only the performance to remove nitrogen or phosphorus from the side streams being treated, but also in its ability to work continually through the flow ranges encountered in the facility during the year.

In many of the processes listed, the dilution of the nitrifier population or the decrease in temperature caused by a larger than typical flow being added to the system will necessitate isolation of the process. The implementation of a protocol of operation for the side stream system with delineated steps to either bypass a portion of the flow being fed to the process or the complete isolation of it when the dilution and decrease of temperature is undesirable.

Another strategy to allow for continual operation of the side stream process is the implementation of equalization, which, given the smaller flows treated in these facilities, should be an option that can be realized without excessive costs or area requirements.

5.0 CASE STUDY—HIGH-RATE CHEMICALLY ENHANCED PRIMARY TREATMENT AS A TOOL FOR WET WEATHER FACILITY OPTIMIZATION

During the design of the Brightwater (Wastewater) Treatment Plant in King County, Washington, the county decided to investigate the use of CEPT for peak flow management to minimize their investment in membrane bioreactors for secondary treatment. This led to full-scale testing conducted at the East Section Reclamation Plant ("South Plant") in Seattle, Washington, over the winter of 2004–2005.

The initial results from jar testing were promising, with up to 56% removal of BOD and 88% removal of TSS at ferric chloride doses of 60 mg/L and anionic polymer doses of 0.5 mg/L. Subsequent jar testing with a blend of ferric chloride and PAC, followed by polymer, produced a notably higher percent removal.

Based on these preliminary jar test results, the full-scale testing proceeded using staged coagulation with ferric chloride (20 to 22 mg/L as Fe), followed by a PAC liquid solution dosage of 12 to 15 mg/L, followed by an anionic polymer (0.4 to 0.6 mg/L).

Raw wastewater characteristics during the storm events recorded during 2001 to 2002 were as follows:

TSS: 178 to 322 mg/L,
BOD: 138 to 283 mg/L, and
COD: 370 to 720 mg/L.

Average particulate BOD fractions ranged from approximately 57 to 69%, with an overall average of 62%. The test showed that the particulate BOD as well as a part of the colloidal BOD were removed.

At peak surface overflow rates (SORs) of 6 m/h (3600 gpd/sq ft), removals were TSS 80 to 95% and BOD 58 to 68%. Removals without chemicals were TSS 50% and BOD 25%.

Removals obtained with SOR as high as 8.8 m/h (5200 gpd/sq ft) were TSS 65 to 80% and BOD 40 to 50%.

The best results were obtained when flocculation time after polymer addition was maximized. Depending on the flows, this time varied from 2 to 6 minutes between the polymer addition point to the primary clarifiers.

6.0 CONCLUSIONS

Results from full-scale pilot testing using site-specific coagulant blends and feed protocols have demonstrated the ability to achieve surface overflow rates of 6 to 6.7 m/h (3600 to 4000 gpd/sq ft) and higher SORs at King County's South Plant.

Chemically enhanced primary treatment is a process widely applicable to high-flow treatment schemes, particularly where flow blending or high flow or load peak shaving within a WRRF are desired. The applicability is particularly promising given the ability to apply CEPT to existing conventional primary clarifiers with appropriate design to promote and maintain flocculation (Krugel et al., 2005; Melcer et al., 2005).

7.0 REFERENCES

Barco, J.; Papiri, S.; Stenstrom, M. K. (2008) First Flush in a Combined Sewer System. *Chemosphere*, **71** (5), 827–833.

Constantine, T.; Katehis, D. (2007) Identifying the Perfect Fit—Deploying the Right Sidestream Nitrogen Removal Process. *Nutrient Removal 2007: State of the Art. WEF/IWA Specialty Conference;* Baltimore, Maryland, March 4–7; Water Environment Federation: Alexandria, Virginia.

Dold, P. L. (2007) Quantifying Sludge Production in Municipal Treatment Plants. *Proceedings of the Water Environment Federation 80th Annual Technical Exhibition and Conference*; San Diego, California, Oct 13–17; Water Environment Federation: Alexandria, Virginia.

Droste, R. L. (1997) *Theory and Practice of Water and Wastewater Treatment;* Wiley & Sons: New York.

Kozak, J.; Patel, K.; Abedin, Z.; Lordi, D.; O'Connor, C.; Granato, T. C.; Kollias, L. (2011) Effect of Ferric Chloride Addition and Holding Time on Gravity Belt Thickening of Waste Activated Sludge. *Water Environ. Res.*, **83**, 140–146.

Krugel, S.; Melcer, H.; Hummel, S.; Butler, R. (2005) High Rate Chemically Enhanced Primary Treatment as a Tool for Wet Weather Plant Optimization and Re-Rating. *Proceedings of the Water Environment Federation 78th Annual Technical Exhibition and Conference;* Washington, D.C., Oct 30–Nov 2; Water Environment Federation: Alexandria, Virginia.

Law, K.; Stinson, B. (2012) Sidestream Treatment Overview. *Proceedings of the IWEA/ISAWWA Joint Annual Conference WATERCON 2012*; Springfield, Illinois, March 19–22; IWEA West: Chicago, Illinois.

Melcer, H.; Krugel, S.; Butler, R.; Carter, P. (2005) Alternative Operational Strategies to Control Pollutants in Peak Wet Weather Flows. *Proceedings of the Water Environment Federation 78th Annual Technical Exhibition and Conference;* Washington, D.C., Oct 30–Nov 2; Water Environment Federation: Alexandria, Virginia.

Metcalf & Eddy, Inc. (2003) *Wastewater Engineering: Treatment and Reuse*, 4th ed.; Tchobanoglous, G., Burton, F. L., Stensel, H. D., Eds.; McGraw-Hill: New York.

National Biosolids Partnership (2005) *National Manual of Good Practice for Biosolids*; National Biosolids Partnership Home Page. http://www.biosolids.org (accessed May 2014).

Osei, K.; Andoh, R. Y. G. (2008) Optimal Grit Removal and Control in Collection Systems and at Treatment Plants. *Proceedings of the World Environmental and Water Resources Congress*; Honolulu, Hawaii, May 12–16; American Society of Civil Engineers: Reston, Virginia.

U.S. Environmental Protection Agency (1994) *Combined Sewer Overflow Control Policy. Fed. Regist*, **59** (75).

U.S. Environmental Protection Agency (1979) *Process Design Manual for Sludge Treatment and Disposal;* EPA-625/1-79-011; U.S. Environmental Protection Agency: Washington, D.C.

Water Environment Federation; American Society of Civil Engineers (1992) *Design of Municipal Wastewater Treatment Plants*, 3rd ed.; WEF Manual of Practice No. 8; ASCE Manuals and Report of Engineering Practice No. 76; Water Environment Federation: Alexandria, Virginia; American Society of Civil Engineers: New York.

Water Environment Federation; American Society of Civil Engineers; Environmental & Water Resources Institute (2010) *Design of Municipal Wastewater Treatment Plants,* 5th ed.; WEF Manual of Practice No. 8; ASCE Manuals and Reports on Engineering Practice No. 76; Water Environment Federation: Alexandria, Virginia.

Wilson, G. E. (1998) Why Do Conventional Grit Systems Have Performance Problems? *65th Pacific Northwest Pollution Control Association Annual Conference*; Portland, Oregon, Oct 26–28. Pacific Northwest Pollution Control Association: Hansen, Idaho.

Index

CPSIA information can be obtained
at www.ICGtesting.com
Printed in the USA
LVHW060417030720
659558LV00001B/9